Springer Theses

Recognizing Outstanding Ph.D. Research

Aims and Scope

The series "Springer Theses" brings together a selection of the very best Ph.D. theses from around the world and across the physical sciences. Each volume has been selected for its scientific excellence by an internationally renowned institute or department. For greater accessibility to non-specialists, the published versions include an extended introduction, as well as a foreword by the student's supervisor explaining the special relevance of the work for the field. The series as a whole provides a valuable resource both for newcomers to the research fields, and for other scientists seeking detailed background information on special questions. Finally, it serves as an accredited documentation of the valuable contributions made by today's younger generation of scientists.

How to Apply

Theses prepared at internationally leading universities and institutes may be nominated for publication by supervisors with the endorsement of the head of department. Each such department is entitled to nominate one thesis per year. Enquiries and requests for a nomination form should be sent, together with the thesis, to Ms. Nathalie Berg (nathalie.berg@springer.com).

Eligibility Criteria

To qualify for consideration, theses should fulfil the following criteria:

- They must be written in good English.
- The topic should fall within the confines of Chemistry, Physics, Earth Sciences, Engineering and related interdisciplinary fields such as Materials, Nanoscience, Chemical Engineering, Complex Systems and Biophysics.
- The work reported in the thesis must represent a significant scientific advance.
- A maximum of 30% of the text may be a verbatim reproduction from the author's previous publications.
- They must have been examined and passed during the 12 months prior to nomination.
- The theses should have a clearly defined structure including an introduction accessible to new PhD students and scientists not expert in the relevant field.
- After acceptance, theses should be supplemented by a "Supervisor's Foreword" outlining the wider scientific relevance of the work.

Indexed by zbMATH.

Sergio Gonzalez-Munoz

Interplay of Nanoscale Physical Properties in 2D Materials and Heterostructures

Springer

Author
Dr. Sergio Gonzalez-Munoz
School of Materials Science
and Engineering
Georgia Institute of Technology
Atlanta, USA

Supervisor
Prof. Oleg V. Kolosov
Physics Department
Lancaster University
Lancaster, UK

ISSN 2190-5053 ISSN 2190-5061 (electronic)
Springer Theses
ISBN 978-3-032-14270-2 ISBN 978-3-032-14271-9 (eBook)
https://doi.org/10.1007/978-3-032-14271-9

This Springer imprint is published by the registered company Springer Nature Switzerland AG
The registered company address is: Gewerbestrasse 11, 6330 Cham, Switzerland

Supervisor's Foreword

The unique physical properties of 2D materials, also called van der Waals (vdW) materials due to the vdW forces that bond their atomically thin layers, are yet to be fully explored, significantly hindering their applications. The mechanical and thermal characteristics of the extensive 2D materials family are much less explored compared to their electronic properties, however, they include record breaking heat conductivity, key for their use as heat spreaders in electronics and thermoelectrics, and outstanding highest known mechanical strength, defining the stability of 2D materials and devices. The challenge in exploring these properties lies in the relative lack of existing instrumentation and methodology capable of quantifying nanoscale mechanical properties and thermal transport in atomically thin layers with micro-range lateral size. Prior to Sergio Gonzalez-Munoz's work, scanning probe microscopy (SPM) was explored by different groups worldwide, including our group in Lancaster, as the best candidate for such studies. Unfortunately, such research merely provided qualitative maps of the thermomechanical behaviour in 2D materials and their heterostructures. In his Ph.D. project, Sergio targeted this challenge, developing innovative SPM approaches and advanced modelling to quantify three fundamental classes of nanoscale physical properties of 2D materials—thermal, mechanical and electrical. Furthermore, the novel research instruments and methods he developed, allowed to reveal cross-interactions between these fundamental physical properties, i.e., thermoelectric, electromechanical and thermomechanical, revealing novel phenomena described in detail in this thesis. The development of direct quantitative measurements of anisotropic thermal properties of vdW nanolayers enabled the discovery of a 2D heterostructure with record thermoelectric efficiency and opened pathways for improving stability of perovskite solar cells. The pioneering studies of the local thermal properties of carbon and hexagonal boron nitride nanotube buckypapers described here pave the way for advanced thermal interface materials, whereas measurements of the local mechanical moduli of twisted graphene bilayers using ultrasonic force microscopy uncovered the complex electromechanics of domains in such structures. Building on the recent discovery of geometric thermoelectricity in nanoscale junctions that locally modifies the Seebeck coefficient, Sergio combined the experimental studies and advanced finite elements simulations to create new artificial thermoelectric nanopatterned materials. This thesis summarises major developments in the physical science and technology of 2D materials and their heterostructures, made possible by Sergio's unique qualities of inquisitive experimentalist and dedicated analytical theorist. Sergio Gonzalez-Munoz's book provides essential precise details not available elsewhere, to benefit researchers and students working in the new and rapidly expanding 2D materials research area.

Lancaster, UK
August 2025

Prof. Oleg V. Kolosov

Abstract

Modern technology has reached an unprecedented point enabling the creation of stable singular atomic layer structures known as two-dimensional materials. These layered systems represent an accessible platform for a deep understanding of the nanoscale physical phenomena, ranging from the fundamental physical fields up to more intricate paradigms like thermoelectricity or electromechanics. The properties of these materials can also be extensively modified, e.g., by stacking layers forming heterostructures or introducing defects that filter the charge transport. Nevertheless, to access such nanoscale systems and quantify their characteristics, powerful and precise tools are needed. Scanning probe microscopes can effectively accomplish this objective, given their capacity to probe a broad spectrum of physical phenomena with enhanced resolution while avoiding the modification of the studied materials. However, they are usually accompanied by analytical methods to deconvolute the intrinsic parameters that describe the structure's performance.

This thesis provides novel pathways for an accessible analysis of the thermal, electronic and mechanical behaviour in low-dimensional materials, focusing on local probe characterisation. Also, the interplay of such properties is investigated, yielding deeper insights into the fields of thermoelectricity, piezoelectricity and thermomechanics. Specifically, the anisotropic thermal behaviour of indium selenide and perovskite is probed with a novel scanning thermal microscopy methodology, which is also applied to quantify the thermoelectric efficiency of structures based on transition metal dichalcogenides. Local modulation of the thermovoltage is also achieved by the patterning of nanoconstrictions in two-dimensional layers. Finally, the electromechanical and thermomechanical response of nanoscale systems is probed in a Moiré superlattice and soft nanotube networks.

This work represents an improvement for the previously developed local probe characterisation techniques, extensively describing new routes for enhanced physical phenomena quantification. It also provides new insights into the behaviour of nanoscale structures, expanding the current knowledge in fields such as heat management, thermoelectricity and piezoelectricity.

List of Publications

Peer-Reviewed Journal Articles

- Ahmad, M., Agarwal, K., Gonzalez-Munoz, S., Ghosh, A., Kodan, N., Kolosov, O. V. & Mehta, B. R. Engineering interfacial effects in electron and phonon transport of Sb_2Te_3/MoS_2 multilayer for thermoelectric ZT above 2.0. *Adv. Funct. Mater.* **32**, 2206384 (2022).
- Maiti, A., Agarwal, K., Gonzalez-Munoz, S. & Kolosov, O. V. Quantifying anisotropic thermal transport in two-dimensional perovskite (PEA_2PbI_4) through cross-sectional scanning thermal microscopy. *Phys. Rev. Mater.* **7**, 023801 (2023).
- Chen, Y., Wu, W., Gonzalez-Munoz, S., Forcieri, L., Wells, C., Jarvis, S. P., Wu, F., Young, R., Dey, A., Isaacs, M., Nagarathinam, M., Palgrave, R. G., Tapia-Ruiz, N. & Kolosov, O. V. Nanoarchitecture factors of solid electrolyte interphase formation via 3D nano-rheology microscopy and surface force-distance spectroscopy. *Nat. Commun.* **14**, 1321 (2023).
- Canetta, A., Gonzalez-Munoz, S., Nguyen, V.-H., Agarwal, K., de Crombrugghe de Picquendaele, P., Hong, Y., Mohapatra, S., Watanabe, K., Taniguchi, T., Nysten, B., Hackens, B., Ribeiro-Palau, R., Charlier, J.-C., Kolosov, O. V., Spièce, J. & Gehring, P. Quantifying the local mechanical properties of twisted double bilayer graphene. *Nanoscale* **15**, 8134–8140 (2023). **Highlighted on the Inside front cover of the issue**.
- Gonzalez-Munoz, S., Agarwal, K., Castanon, E. G., Kudrynskyi, Z. R., Kovalyuk, Z. D., Spièce, J., Kazakova, O., Patanè, A. & Kolosov, O. V. Direct measurements of anisotropic thermal transport in γ-InSe nanolayers via cross-sectional scanning thermal microscopy. *Adv. Mater. Interfaces* **10**, 2300081

(2023). **Highlighted on the Cover Picture and the Editor's Choice of the issue**.

- Jones, R. S., Gonzalez-Munoz, S., Griffiths, I., Holdway, P., Evers, K., Luanwuthi, S., Maciejewska, B. M., Kolosov, O. & Grobert, N. Thermal conductivity of carbon/boron nitride heteronanotube and boron nitride nanotube buckypapers: implications for thermal management composites. *ACS Appl. Nano Mater.* **6**, 15374–15384 (2023). **Highlighted on the Cover of the issue**.

Preprints

- Ghosh, A., Agarwal, K., Gonzalez-Munoz, S., Bisht, P., Vishwakarma, C. K., Kaur, N., Ahmad, M., Vullum, P. E., Belle, B. D., Singh, R., Kolosov, O. V. & Mehta, B. R. Tuning thermoelectric properties of Sb_2Te_3-$AgSbTe_2$ nanocomposite thin film—synergy of band engineering and heat transport modulation. Preprint at https://arxiv.org/abs/2305.19217 (2023).
- Canetta, A., Gonzalez-Munoz, S., Nguyen, V.-H., Agarwal, K., de Crombrugghe de Picquendaele, P., Hong, Y., Mohapatra, S., Watanabe, K., Taniguchi, T., Nysten, B., Hackens, B., Ribeiro-Palau, R., Charlier, J.-C., Kolosov, O. V., Spièce, J. & Gehring, P. Quantifying the local mechanical properties of twisted double bilayer graphene. Preprint at https://arxiv.org/abs/2405.12610 (2024).

Journal Articles Submitted

- Ghosh, A., Agarwal, K., Gonzalez-Munoz, S., Bisht, P., Vishwakarma, C. K., Kaur, N., Ahmad, M., Vullum, P. E., Belle, B. D., Singh, R., Kolosov, O. V. & Mehta, B. R. Enhancing thermoelectric properties of Sb_2Te_3-$AgSbTe_2$ nanocomposite thin film—synergy of band engineering and heat transport modulation (Submitted to *Adv. Sci.*).

Journal Article in Preparation

- Gonzalez-Munoz, S., Castanon, E. G., Gorbachev, R. V., Niblett, A., Agarwal, K., Kazakova, O., Ponomarenko, L. A., McCann, E., Fal'ko, V. I. & Kolosov, O. V. Symmetry effects in the single 2D material geometrical thermoelectric phenomena.
- Canetta, A., Gonzalez-Munoz, S., Agarwal, K., de Crombrugghe, P., Nguyen, V.-H., Hong, Y., Mohapatra, S., Watanabe, K., Taniguchi, T., Nysten, B., Hackens, B., Ribeiro-Palau, R., Charlier, J.-C., Kolosov, O., Spièce, J. & Gehring, P. Local electromechanical properties of twisted double bilayer graphene.

Posters, Oral Presentations and Outreach

Parts of this work have been presented at the following scientific conferences as a main presenter (see Sect. A.1 for other contributed presentations):

- Agarwal, K., Gonzalez-Munoz, S., Castanon, E., Niblett, A., Kudrynskyi, Z. R., Kazakova, O., Patanè, A. & Kolosov, O. V. Scanning thermal microscopy of 2D materials in high vacuum environment. *Microscience Microscopy Congress (MMC) 2021* (Virtual). **Poster and flash talk presentation**.
- Gonzalez-Munoz, S., Agarwal, K., Castanon, E., Kudrynskyi, Z., Kovalyuk, Z. D., Kazakova, O., Patanè, A. & Kolosov, O. Measurements of nanoscale thermal transport and its anisotropy in vdW materials via cross-sectional scanning thermal microscopy (xSThM). *Graphene Week 2021* (Virtual). **Poster and flash talk presentation**.
- Gonzalez-Munoz, S., Agarwal, K., Castanon, E., Kudrynskyi, Z., Kovalyuk, Z., Spièce, J., Kazakova, O., Patanè, A. & Kolosov, O. Quantitative measurements of anisotropic thermal transport in gamma indium selenide via cross-sectional scanning thermal microscopy. *FST Science Week 2022* (Lancaster, United Kingdom). **Poster presentation**.
- Gonzalez-Munoz, S., Agarwal, K., Castanon, E., Kudrynskyi, Z., Kovalyuk, Z., Spièce, J., Kazakova, O., Patanè, A. & Kolosov, O. Quantitative measurements of anisotropic thermal transport in vdW materials via cross-sectional scanning thermal microscopy (xSThM). *2022 MRS Spring Meeting & Exhibit* (Honolulu, United States). **Oral presentation**.
- Gonzalez-Munoz, S., Agarwal, K., Castanon, E., Kudrynskyi, Z., Kovalyuk, Z., Spièce, J., Kazakova, O., Patanè, A. & Kolosov, O. Quantitative measurements of anisotropic thermal transport in vdW materials via cross-sectional scanning thermal microscopy (xSThM). *E-MRS 2022 Spring Meeting* (Virtual). **Oral presentation**.
- Gonzalez-Munoz, S., Agarwal, K., Castanon, E., Kudrynskyi, Z., Kovalyuk, Z., Spièce, J., Kazakova, O., Patanè, A. & Kolosov, O. Quantitative measurements of anisotropic thermal transport in gamma indium selenide via cross-sectional scanning thermal microscopy. *Condensed Matter and Quantum Materials 2022* (Bath, United Kingdom). **Oral presentation**.
- Gonzalez-Munoz, S., Agarwal, K., Castanon, E., Kudrynskyi, Z., Kovalyuk, Z., Spièce, J., Kazakova, O., Patanè, A. & Kolosov, O. Measurements of anisotropic thermal transport in γ-InSe via quantitative cross-sectional scanning thermal microscopy. *Virtual RMS AFM & SPM Meeting 2022* (Virtual). **Submitted talk**.

- Gonzalez-Munoz, S., Agarwal, K., Castanon, E. G., Kudrynskyi, Z. R., Kovalyuk, Z. D., Spièce, J., Kazakova, O. L., Patanè, A. & Kolosov, O. V. Measurements of anisotropic thermal transport in γ-InSe via quantitative cross-sectional scanning thermal microscopy. *Graphene Week 2022* (Munich, Germany). **Poster presentation**.
- Gonzalez-Munoz, S., Agarwal, K., Ghosh, A., Ahmad, M., Mehta, B. R. & Kolosov, O. V. Anisotropic thermal transport using xSThM studies in 2D-3D heterostructures and composites. *Graphene Week 2022* (Munich, Germany). **Poster presentation**.
- Gonzalez-Munoz, S., Jones, R. S., Maciejewska, B., Grobert, N. & Kolosov, O. V. Thermomechanical study via piercing indentation SThM in C/BN heteronanotubes buckypapers. *Microscience Microscopy Congress (MMC) 2023* (Manchester, United Kingdom). **Contributed talk**.
- Gonzalez-Munoz, S., Agarwal, K., Maiti, A. & Kolosov, O. V. Quantitative scanning thermal microscopy studies of the influence of interfaces and heat transport anisotropy in 2D materials. *Microscience Microscopy Congress (MMC) 2023* (Manchester, United Kingdom). **Poster presentation**.
- Gonzalez-Munoz, S., Canetta, A., Agarwal, K., Spièce, J., Gehring, P. & Kolosov, O. V. Correlation of nanoscale electromechanical and mechanical properties of twisted double bi-layer graphene via UFM, PFM, and E-HFM. *Microscience Microscopy Congress (MMC) 2023* (Manchester, United Kingdom). **Poster presentation**.
- Gonzalez-Munoz, S., Agarwal, K., Maiti, A. & Kolosov, O. V. Quantitative scanning thermal microscopy studies of the influence of interfaces and heat transport anisotropy in 2D materials. *Graphene Week 2023* (Gothenburg, Sweden). **Poster presentation**.
- Gonzalez-Munoz, S. Interplay of nanoscale physical properties in 2D materials and heterostructures—three minute thesis competition. *E-MRS 2023 Fall Meeting* (Warsaw, Poland). **Contributed talk**.
- Gonzalez-Munoz, S., Agarwal, K., Maiti, A. & Kolosov, O. V. Quantitative scanning thermal microscopy studies of the influence of interfaces and heat transport anisotropy in 2D materials. *E-MRS 2023 Fall Meeting* (Warsaw, Poland). **Contributed talk**.
- Gonzalez-Munoz, S. Quantitative probing of nanoscale heat transport via advanced scanning thermal microscopy. *National Institute of Standards and Technology* (Virtual). **Invited talk**.

- Gonzalez-Munoz, S., Agarwal, K. & Kolosov, O. V. Enhanced thermoelectricity and thermal anisotropy of 2D-based structures. *AFM & SPM Meeting 2024* (Durham, United Kingdom). **Poster presentation**.
- Gonzalez-Munoz, S. Probing the nanophysics interplay via advanced scanning probe microscopy. *Ludwig Maximilian University of Munich* (Munich, Germany). **Invited talk**.

Parts of this work have been recognised with the following awards and grants:
Awards:

- 2023 departmental prize for outstanding research and contributions to high-impact publications. *Physics Department, Lancaster University.*
- Young researcher award. *E-MRS 2023 Fall Meeting, European Materials Research Society.*

Grants:

- 2022 Graphene Week student grant. *Graphene Flagship, European Union.*
- 2022 research student conference fund. *Institute of Physics, United Kingdom.*
- 2023 Graphene Week student grant. *Graphene Flagship, European Union.*

Acknowledgements Undertaking a Ph.D. program in physics is a path with plenty of challenges, but, at the same time, it is a unique opportunity to expand your knowledge frontiers to the highest level, debut as a young professional scientist, and, most importantly, create a plethora of wonderful memories. During the four years of my Ph.D., I have been supported and guided by many people, without which I would not have been able to finish this project. In this section, I would like to express my sincere appreciation to all the people who have helped and influenced me.

Firstly, I would like to thank my supervisor, Prof. Oleg V. Kolosov, for always providing encouragement and support. Since the beginning of my Ph.D., he has been there resolving my questions with his vast physics knowledge, helping with the experiments, discussing the results and giving a hand in the laboratories whenever I needed him. Not only this, but he has also supported and advised me on many obstacles inside and outside the Ph.D., watching me grow into a better scientist and a better person, for which I am extremely grateful.

I am also thankful to the Lancaster University Physics Department for providing me with the financial scholarship, logistic resources, and advice for the Ph.D. postgraduate program. The Lancaster University Physics Department is also home to great professional staff, which have made my Ph.D. research more comfortable by providing different resources always at the topmost standards. Amongst them, I would like to acknowledge the mechanical technicians Reece Grimshaw and Aaron Nicholas for manufacturing all the needed components with such high precision. I also appreciate the work done by the electronic workshop personnel (Ashley and Lucy Wilson) since they created magnificent electronic circuits and components vital for many experiments. Subsequentially, the IT support related to software and hardware provided by Robin Lewsey and Rupert Prudom has been fundamental for the smooth development of my research studies. Also, I would like to thank Dr. Kunal Lulla, Garry Vernon and Dr. Yuezhen Lu for their mentorship on the cleanroom fabrication processes. Furthermore, every visit to the Physics store to collect packages has been a superb opportunity to have a nice chat with Stephen Holden and Grainne Wilkinson.

The scientific environment of the Lancaster University Physics Department has allowed me to meet fantastic colleagues, each in different career places but with a shared love for science. I would like to mention a few of these people with whom I have shared my research period and who have significantly influenced my Ph.D. and personal life. I must acknowledge the current and former members of the Quantum Nanotechnology Kolosov's group. I would like to express my gratitude to Dr. Khushboo Agarwal for the uncountable hours spent working in the labs and for introducing me to Indian culture with delicious food and fantastic festivity events. To Dr. Rajasekhar Medapalli, for introducing me to cricket and for all the funny moments in C28 talking about the US, India and Spain. To Dr. Yue Chen, for being a great friend who has shown and taught me so many important life values, scientific perspectives and Chinese culture. And to Dr. Eli G. Castanon for helping me with my first Ph.D. steps and for all the Asturian-Catalan "ieepaaa" and "ese pa ahí" greetings. Outside of the group, I would also like to recognise Fangling Wu, an enthusiastic person who distracted me from my work but was always there to have profound chats about life.

External to the scientific context, I must mention specific people who, without their support, I would not have been able to complete my Ph.D. journey. The central encouragement during these years has come from my family, especially from my parents, Ana and Luis. They have been my life role models, being by my side whenever and wherever I needed them, teaching me infinite wise advice that made me grow to be a nice and capable person and being always the first ones to pick me up whenever I fall and feel proud and happy whenever I succeed. I am also especially grateful and proud of my brother Raúl, who has also been an amazing emotional support during these years, visiting me in Lancaster, always being there to talk, sharing meaningful tattoos and promoting my active side. I also want to acknowledge my cousin Marc for the invaluable support when I needed it the most, for his visits to Lancaster and for sharing my view of life, which showed me what the real meaning of "having someone's back" means. Moreover, I truly appreciate the support of my grandmother Yaya Maria, who learnt how to videocall me at 89 years of age and encouraged my overseas period. Lastly, I want to thank my friends, Anna, Fonta and Xavi for giving me the wings necessary to undertake this project and for their continuous advice, support and care.

Competing Interests The author has no competing interests to declare that are relevant to the content of this manuscript.

Contents

Abbreviations

1LG	Monolayer graphene
2D	Two-dimensional
A-C	Armchair
AFM	Atomic force microscopy
BEXP	Beam exit cross-sectional polishing
BNNT	Boron nitride nanotube
BP	Buckypaper
CNT	Carbon nanotube
DIW	Deionised water
DOS	Density of states
DSP	Double saddle point
E-HFM	Electrical heterodyne force microscopy
FEA	Finite element analysis
FLG	Few-layer graphene
HV	High vacuum
IPA	Isopropanol
KPFM	Kelvin probe force microscopy
LIA	Lock-in amplifier
LT	Low temperature
MW	Multiwall
PDMS	Poly(dimethylsiloxane)
PFM	Piezoresponse force microscopy
PMMA	Poly(methyl methacrylate)
PPC	Poly(propylene carbonate)
RT	Room temperature
SEM	Scanning electron microscopy
SPL	Scanning probe lithography
SPM	Scanning probe microscopy
STGM	Scanning thermal gate microscopy
SThM	Scanning thermal microscopy
TDBG	Twisted double bilayer graphene

TE	Thermoelectricity
TEM	Transmission electron microscopy
TMD	Transition metal dichalcogenide
UFM	Ultrasonic force microscopy
vdW	van der Waals
Z-Z	Zigzag

Symbols, Variables and Constants

β	Lattice constant [m]
Γ_S	Shear modulus [Pa]
δ	Interlayer distance [m]
ε	Strain
η	Efficiency [%]
Θ	Heat current [W]
θ	Angle [°]
κ	Thermal conductivity [$Wm^{-1}K^{-1}$]
λ	Mean free path [m]
μ	Charge carrier mobility [$m^2V^{-1}s^{-1}$]
ν	Poisson's ratio
σ	Electrical conductivity [Sm^{-1}]
τ_T	Tensile strength [Pa]
Φ	Heat flux [Wm^{-2}]
φ	Thermal conductivity ratio
ψ	Roughness [m]
ω	Frequency [Hz]
A	Amplitude [m]
a	Radius [m]
C	Capacitance [F]
E	Young's modulus [Pa]
F	Force [N]
G	Thermal conductance [WK^{-1}]
h	Thickness [m] or Planck's constant [Js]
I	Current [A]
J	Bessel function of 1st kind
K	Stiffness [Nm^{-1}]
k_B	Boltzmann's constant [JK^{-1}]
l	Length [m]
m	Mass [kg]
n	Charge carrier density [m^{-3}]

P	Power [W]
p	Pressure [Pa]
q	Electric charge [C]
R	Thermal resistance [KW^{-1}]
r	Thermal insulance [Km^2W^{-1}]
S	Seebeck coefficient [VK^{-1}]
s	Stress [Pa]
T	Temperature [K]
t	Time [s]
V	Voltage [V]
W	Energy [eV]
w	Width [m]
x	Position [m]
Z	Separation distance [m]
ZT	Figure of Merit

List of Figures

List of Tables

Chapter 1
Introduction

Abstract The opening chapter establishes the scientific motivation behind this research by illustrating how nanoscale modifications in solid-state systems can trigger complex, interconnected changes in their physical behaviour. This interplay is especially significant in low-dimensional materials, where quantum effects dominate. A central focus is placed on two-dimensional (2D) materials due to their tunable, anisotropic properties and their importance in applications requiring miniaturisation and high performance. Beyond traditional fields like electronics and photonics, these materials also offer promising capabilities in thermomechanics and nanomechanics. The chapter emphasises the need to understand both individual transport properties and their coupling, reflected in phenomena such as thermoelectricity and electromechanics. Two foundational aspects of this research are introduced: the controlled fabrication of 2D materials to preserve their intrinsic properties and their high-resolution, minimally invasive characterisation. Scanning probe microscopy (SPM) is presented as a key tool, offering quantitative, local analysis of nanoscale physical parameters essential for exploring multiphysics interactions in 2D systems. In the end of the chapter, a brief structure of the contents of this book is provided to the reader.

Similarly to the chain reaction caused by the disturbance of an initial piece in a domino toppling, the modification of an individual element in a solid-state material could result in related perturbations on many other building blocks of the entire structure, changing the fundamental physical properties, i.e., mechanical, optical, thermal, electrical, etc. and their conjunct behaviour or interplay. For example, electrons can simultaneously contribute to thermal and electric transport and be influential in the mechanical stability of the system. This linked phenomenon is accentuated in materials with nanoscale dimensions, where quantum physics assumes the principal role of their behaviour. Hence, cross-interaction fields such as thermoelectrics or electromechanics are gaining increasing importance in the low-dimensional structures' current research.

Specifically, two-dimensional (2D) materials, also known as van der Waals (vdW) structures, are one of the most studied platforms inside the low dimension range. Their

S. Gonzalez-Munoz, *Interplay of Nanoscale Physical Properties in 2D Materials and Heterostructures*, Springer Theses, https://doi.org/10.1007/978-3-032-14271-9_1

intrinsic minimal thickness and different binding configurations lead to varied properties when measured along their basal plane or through it, showing a clear anisotropic degree. Also, they possess a great extent of tunability, with multiple tools that can be used to modulate their features in relatively straightforward ways. For these reasons, 2D materials have been applied to an extensive number of technological applications that require miniaturisation, enhanced performance and in situ manipulation. Furtherly, their investigation has become fundamental in many laboratories around the globe, focusing on aspects such as their optical, electronic and magnetic capabilities. However, the catalogue of features at which vdW structures excel is not limited to just the aforementioned. Other physical fields like heat transport or mechanical performance are just as important. Many 2D materials are characterised by having strong intralayer bonds, promoting their mechanical stability and duration if they are applied in technologies requiring high resistance to stress. On the other hand, the electronic industry is shifting towards high efficiency devices, which are in need of a continuous upgrade of the energy conversion limit at the expense of thermal losses. Here is where thermal management and heat reutilisation can be extremely useful, even more if controlled with the nanoscale precision that two-dimensional structures are capable of reaching.

As mentioned before, physical properties will always affect each other, hence, it is crucial to not only understand the individual fundaments of, e.g., heat and electrical transport but also their interplay, known as thermoelectricity. The same can be applied to systems under mechanical stress subjected simultaneously to an internal or external charge exchange, a field known as electromechanics. Or, to close the circle, systems that induce a thermal response after a mechanical actuation, i.e., thermomechanics. In all these areas, the versatility of 2D materials could be of great use to develop novel understanding and obtain groundbreaking applicable solutions (Fig. 1.1).

The line of advancement is therefore set, however, two main aspects must be especially considered. The first one is the fabrication of such type of nanoscale layers, which, due to their handling difficulty, regularly require their interaction with other supporting structures. Before thinking about any possible application, such fabrication has to be originally performed with the aim of understanding the nature of the isolated vdW structure, therefore minimising its interaction with external systems or inputs. Once this is completed, other environments can be added to the ensemble in a controlled way, forming, for instance, heterostructures or modifying the lattice configuration of the pristine material. The second important aspect is their characterisation, which, again, has to be conducted so that the material is minimally influenced by its interacting elements, or at least in a preestablished and controlled manner. The majority of classical characterisation methods would not fully serve this purpose, therefore, methodologies of investigation capable of obtaining reliable data directly from the 2D materials without modifying their behaviour are required. Advanced scanning probe microscopies (SPM) are the ideal tools for this, given their ability to quantitatively probe independent physical properties with nanoscale resolution and without detrimental effects on the samples. Like 2D materials, they are also capable of excellent tunability and adaptability, even merging with external

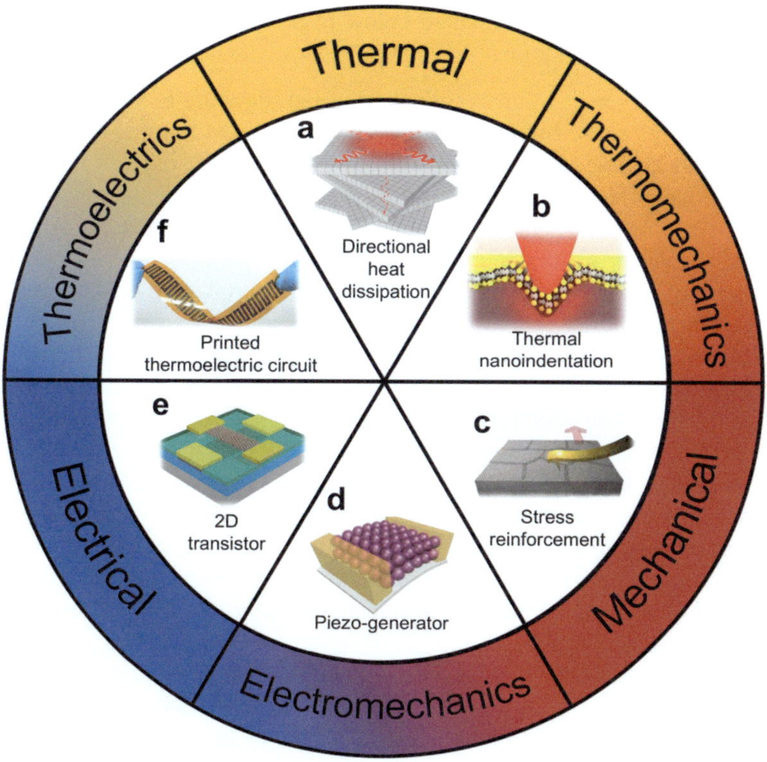

Fig. 1.1 Diagram of the physical properties of 2D materials studied here and their applications. **a** Stacked 2D layers for directional heat dissipation (adapted from Kim et al. (2021), licensed under CC BY 4.0) [1], **b** thermomechanical 2D lithography (adapted from Liu et al. (2020), licensed under CC BY 4.0) [2], **c** 2D material stress reinforcement (adapted from Zheng et al. (2016), licensed under CC BY 4.0) [3], **d** 2D piezoelectric generator (adapted with permission from Ref. [4], Copyright © 2019 Elsevier) [4], **e** a nano-transistor with a 2D channel (adapted from Qian et al. (2019), licensed under CC BY 4.0) [5] and **f** a printed 2D thermoelectric circuit (adapted from Juntunen et al. (2018), licensed under CC BY 4.0) [6]

spectroscopic techniques or functioning under precisely selected conditions or in situ dynamic processes. These perks have been extensively employed to measure the fundamental properties of vdW materials, however, they are rarely the only solution employed to fully obtain a clear idea of the material's behaviour, given that they usually require additional analysis methods or simulations.

In this thesis, accessible characterisation methodologies primarily focused on local probe microscopy are employed to characterise different physical parameters (as well as their cross-linkages) of commonly employed and novel two-dimensional structures. Interesting platforms for the manipulation of such samples are also provided here to serve as prospective high potential solutions in fields such as thermoelectricity, nanomechanics, and heat management. These goals are discussed in the following chapters, but a brief explanation of each of these sections including their aims and objectives is presented below to provide the reader with an initial overview of the described topics:

- Firstly, Chap. 2 serves as an examination of the literature focused on the fundamental principles discussed in this study, with the aim of framing and introducing the previous knowledge in the field. Starting with a description of the characteristics of four different 2D materials, followed by their fabrication and characterisation methodologies, and finishing with unconventional phenomena related to such nanoscale structures.
- Chapter 3 elucidates the synthesis procedures undertaken to create the samples analysed in the thesis. The aim of this section is to expose accessible fabrication steps to achieve tailored 2D samples. Specifically, various mechanical exfoliation methodologies are explained, together with the precise process of heterostructure formation as well as patterning and deposition, targeting objectives such as inner layer access, device fabrication and reduced contamination.
- The insight of heat transport in 2D materials is discussed in Chap. 4, aiming towards directional heat flow at the nanoscale. A novel characterisation methodology is employed with the objective of extracting the quantitative anisotropic thermal conductivities of various semiconductor materials as well as the upper bounds of the interfacial thermal insulances between such layered structures and their substrates.
- The knowledge obtained in Chap. 4 is applied in Chap. 5 together with additional electronic characterisation aiming to provide two potential platforms based on 2D systems for enhanced thermoelectric efficiency at room temperature. Here, the optimisation of both charge and heat transport via interfacial engineering is the main objective.
- Chapter 6 focuses on a crucial thermoelectric parameter, the Seebeck coefficient, with the ultimate purpose of gaining the ability to locally modify it via a series of accessible tools, all of them based on the influential effect of nanoconstrictions. With this, milestones such as nanoscale tunning of the thermovoltage and in situ TE characterisation can be achieved.
- Lastly, Chap. 7 is divided into the electromechanical behaviour of twisted bilayer graphene and the thermomechanics of one-dimensional networks known as buckypapers. The used SPM-based techniques aim to characterise such structures for the first time, with the specific purposes of finding a TDBG Moiré superlattice capable of local flexoelectricity and modified Young's modulus together with an enhanced thermal transport in soft buckypapers with a coaxial nanotube structure.

References

1. Kim, S. E. et al. (2021). Extremely anisotropic van der Waals thermal conductors. *Nature, 597*, 660–665.

2. Liu, X. et al. (2020). Thermomechanical nanostraining of two-dimensional materials. *Nano letters, 20*, 8250–8257.
3. Zheng, X. et al. (2016). Robust ultra-low-friction state of graphene via Moiré superlattice confinement. *Nature communications, 7*, 13204.
4. Han, J. K. et al. (2019). Tunable piezoelectric nanogenerators using flexoelectricity of well-ordered hollow 2D MoS_2 shells arrays for energy harvesting. *Nano Energy, 61*, 471–471.
5. Qian, Q. et al. (2019). 2D materials as semiconducting gate for field-effect transistors with inherent over-voltage protection and boosted ON-current. *npj 2D Materials and Applications, 3*, 24.
6. Juntunen, T. et al. (2018). Inkjet printed large-area flexible few-layer graphene thermoelectrics. *Advanced Functional Materials, 28*, 1800480.

Chapter 2
Foundations of Multiphysics at the Nanoscale

Abstract This chapter presents a comprehensive literature review of the fundamental concepts and technical background that support the research in this thesis, focused on two-dimensional (2D) materials and their multiphysical behaviour at the nanoscale. It begins with an overview of the structural, mechanical, electrical and thermal properties of representative 2D materials, emphasizing their anisotropy, tunability and relevance in next-generation technologies. The review then explores key fabrication strategies, including mechanical exfoliation, chemical vapour deposition and a variety of lithographic and scanning probe-based patterning techniques. Particular attention is given to the integration of electrical contacts, which is essential for enabling functional nanodevices. The chapter proceeds with an examination of characterisation methodologies, ranging from conventional microscopy and spectroscopy to advanced scanning probe techniques used to probe nanomechanical, electrical and thermal responses with high spatial resolution. Finally, the discussion extends to advanced physical phenomena emerging in 2D systems, including directional heat transport, interfacial thermal resistance and coupled behaviours such as thermoelectric and electromechanical effects. By identifying gaps in the current understanding of these phenomena, this chapter defines the motivation and research directions pursued in the experimental chapters that follow.

2.1 Overview

Nanotechnology exploits novel physical phenomena only accessible when reducing certain materials at the nanoscale to conceive ground-breaking applications. Within this emerging but promising nanoworld, the investigation of 2D materials is an arising branch. 2D materials are layered crystalline structures with nanoscale thicknesses several orders of magnitude smaller than their lateral sizes (ratio $\geq$ 1000). These materials are generally arranged as a vdW stack of covalently intrabonded nanosheets known as monolayers, generating novel features that discern from their bulk three-dimensional (3D) counterparts in properties such as the surface energy,

S. Gonzalez-Munoz, *Interplay of Nanoscale Physical Properties in 2D Materials and Heterostructures*, Springer Theses, https://doi.org/10.1007/978-3-032-14271-9_2

particle density, epitaxy or shape. Although the number of layers liming the transition from 2 to 3D properties varies between materials, it is generally accepted that, beyond 10 layers, a structure is not considered 2D anymore [1]. These dimensional differences cause the emergence of unique characteristics in different physical fields.

Graphene is the best-known 2D material on the list, mainly due to its distinctive set of properties. However, it has hindrances to perform in applications that require, e.g., thermal or electrical insulation. To cover the areas in which graphene cannot be used, other 2D materials are investigated. Scientists can customise 2D structures to adapt to the required outcome, e.g., studying 2D transition metal dichalcogenides (TMD) or hexagonal boron nitrides (h-BN) for semiconductor and insulator applications, respectively. Additionally, the efficient production of vdW nanostructures has been a matter of interest due to the lack of quality/cost preparation methods. Conversely, the number of characterisation techniques capable of quantifying such nano-systems is extensive, enabling the precise and multidisciplinary study of several characteristics of the 2D materials.

In the first section of this review, four different 2D materials are described to outline their main physical properties. In the second part, the most employed 2D materials production techniques are elucidated. The third section summarises multiple characterisation methodologies used to study vdW nanostructures. Lastly, physical phenomena related to 2D systems and devices are discussed to provide an insight into their advanced and interlinked features.

2.2 Two-Dimensional Materials and Their Properties

2.2.1 Two-Dimensional Materials

Graphene

Over the years, graphene has become the 2D material par excellence. Essentially, it is a monoatomically thin layer of carbon atoms arranged in a hexagonal honeycomb lattice. Multiple graphene layers can be vertically stacked through van der Waals forces to form graphite, consequently cataloguing both materials as carbon allotropes [2].

Although initial theoretical studies of this material date back to the late 1940s [3], it was not until 2004 when a graphene monolayer was isolated and characterised by the Nobel prize laureates Geim and Novoselov [4]. They laminated a graphite structure via mechanical exfoliation, separating graphene layers of minimal thicknesses and transferring them onto a silicon dioxide (SiO_2) substrate while keeping a quasi-neutral state of the material. This study proved the applicability and efficacy of 2D materials in real-world devices and opened the door to further research conducted on other similar vdW structures [5].

According to the standard nomenclature, the terminology for graphene ranges from a stack of 10 or less graphene layers, namely few-layer graphene (FLG), to

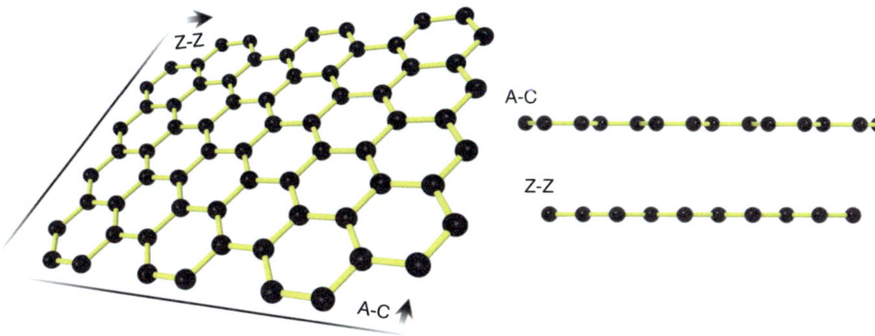

Fig. 2.1 Graphene monolayer structure composed of carbon (black) atoms with the armchair (A-C) and zigzag (Z-Z) edges' views

single- or monolayer graphene (1LG) [6]. In 1LG (Fig. 2.1), each carbon atom is covalently σ-bonded in the horizontal or basal plane to three other neighbouring atoms, forming a lattice with outstanding mechanical and thermal properties [7, 8]. This type of binding is known as sp^2 hybridisation since each atom shares three valence electrons forming a binding angle of 120° [9]. The remaining valence electron is located in the cross-plane (z) vertical direction, being capable of forming vdW interactions with other layers. This perpendicular bonding creates two half-filled bands called π and π^*, which are responsible for the exceptional electrical and optoelectronic properties of graphene [10] and position it as a semi-metallic material [2].

Transition Metal Dichalcogenides

Transition metal dichalcogenides (TMDs) are a family of 2D materials with common morphology and composition features. The stoichiometric formula for these materials is MX_2, being M a transition metal (molybdenum, tungsten, niobium, vanadium, etc.) and X a chalcogen (sulphur, selenium, tellurium, etc.). Some distinguished TMD compositions are MoS_2, $MoSe_2$, WS_2, and WSe_2 [11].

Monolayer TMDs are arranged in a "sandwich" distribution, where an atomically thin layer of transition metals is encapsulated between two chalcogen atomic planes (Fig. 2.2). The internal bonding of these lattices is covalent between the basal elements [12]. These monolayers can be understood as single building blocks that stack vertically through weak vdW forces until they transition from nanoscale to bulk thicknesses, analogous to other 2D materials, allowing easy isolation of TMDs single layers by exfoliation or other production methods [11].

Although having notable mechanical properties due to their interatomic covalent bonds, TMDs are most well-known for their unique optical and electronic behaviour. The different compositions and crystallinities in the atomic layers of these materials cause an array of distinct symmetries that lead to different physical characteristics. For instance, the optical properties of odd-layer TMDs are influenced by their broken inversion symmetry, splitting the valence band and causing, e.g., a

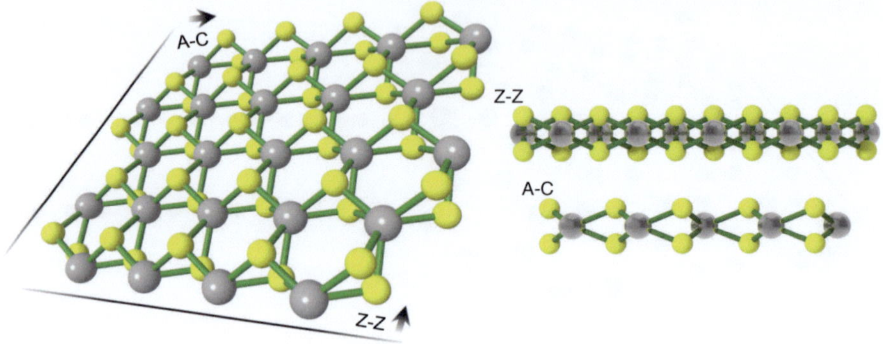

Fig. 2.2 MoS_2 monolayer structure composed of molybdenum (grey) and sulphur (yellow) atoms with the A-C and Z-Z edges' views

valley-selective circular dichroism [13, 14]. Simultaneously, an electronic correlation occurs, changing its electrical properties and the classification of the material as metallic, semiconductor or insulator depending on the transition metal present in the structure [9, 15].

Indium Selenide

Given the proximity of indium to the transition metals group in the periodic table, indium selenide (InSe) is often mistakenly classified as a TMD, nonetheless, this structure is correctly categorised as a post-TMD. This material can be subdivided into two groups depending on their stoichiometry, namely indium (II) selenide (InSe) and indium (III) selenide (In_2Se_3). Each of these subgroups has different phases known as polytypes (β, γ and ε for InSe and α, β, γ, δ and κ for In_2Se_3), but only the α and β polymorphs of In_2Se_3 and all the polytypes of InSe exists with a layered structure [16–18]. Although not considered a TMD, InSe possesses many similar properties, starting from the monolayer arrangements. For the layered In_2Se_3 structures, in the α phase, the indium atoms are located in the tetrahedral cavities formed by four selenium elements, while in the β phase, the indium atoms are situated in the octahedral locations [16]. Conversely, the InSe polytypes (Fig. 2.3) are arranged with hexagonal symmetry in the β and ε phases and rhombohedral symmetry in the γ phase [18, 19]. This material also shares the requirements to be recognised as a 2D structure since it owns intralayer covalent bonds and interlayer vdW forces [20].

This material is well-known for its high electron mobility and characteristic direct bandgap, which, combined with a notable elasticity, position InSe as one strong perspective in the semiconductor field [21]. Furthermore, InSe has a high photoresponse, which makes it suitable for optical applications such as phototransistors and solar cells [16]. Also, its phase heterogeneity yields exciting potential in the thermal and electric fields [20].

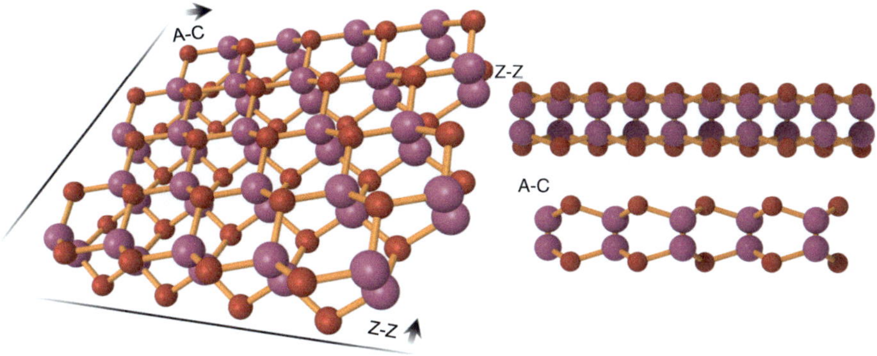

Fig. 2.3 InSe monolayer structure composed of indium (purple) and selenium (red) atoms with the A-C and Z-Z edges' views

Hexagonal Boron Nitride

Similar to graphene, hexagonal boron nitride (h-BN) monolayer is a 2D material with a single-atom-thick basal plane and a hexagonal honeycomb lattice. However, the structure is a compound formed of boron and nitrogen atoms with some very distinct properties from those of graphene.

The boron and nitrogen atoms in the in-plane (xy) direction are covalently bonded so that each boron atom is linked with three nitrogen atoms and vice versa (Fig. 2.4) [22]. Although vdW forces act between the monolayers, their commensurate state is yet not as straightforward. Contrary to graphene, where each carbon atom is vertically interacting with the centre of a benzene ring on the neighbouring layers, in h-BN, the atoms are eclipsed by the counter element in the upper and lower basal planes. Hence, each boron atom is linked to two nitrogen atoms in the vertical direction by vdW interactions, one in the upper lattice and the other in the lower one. The reason behind this is the partially ionic character of h-BN, which forces this stacking configuration to be the most stable one [23, 24].

h-BN distinguishes itself from other materials mainly in the heat transport branch, given its high stability and superior thermal conductivity [25]. Due to the variation in the in-plane and cross-plane interactions, the majority of the properties in this material show elevated anisotropy, a feature shared with the other mentioned 2D materials. Noticeably, h-BN is an electrical insulator, given its high bandgap energy [26–28].

In summary, Table 2.1 shows the interatomic spacing and the monolayer thicknesses of the mentioned 2D materials.

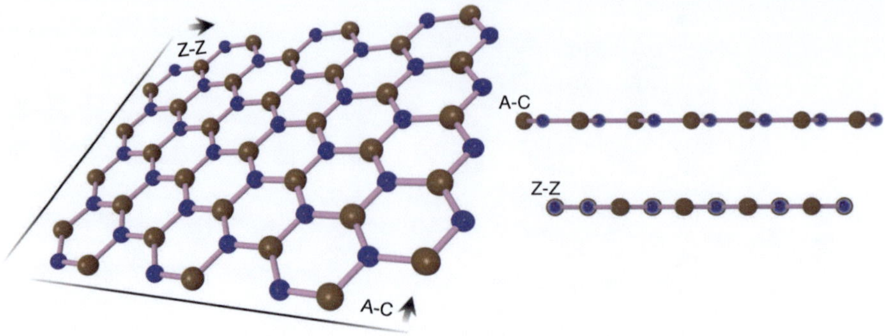

Fig. 2.4 h-BN monolayer structure composed of boron (brown) and nitrogen (blue) atoms with the A-C and Z-Z edges' views

Table 2.1 Thicknesses and interatomic spacings of different 2D materials

Material	Thickness (Å)	Interatomic spacing (Å)	References
Graphene	3.35–3.37	1.42	[3, 7]
MoS_2	6.50	2.41	[29]
α-In_2Se_3	10.20–12	–	[30, 31]
γ-InSe	8.30	4.01	[18]
h-BN	3.33	1.45	[28, 32]

2.2.2 *Nanomechanics*

Young's Modulus

One of the most representative attributes of a material's mechanical performance is the Young's modulus (E), with units of Pa in the SI system. This characteristic results from the coefficient between the tension applied to the material and its consequent deformation, directly relating the stress (s) and strain (ε) on the structure ($E = s\varepsilon^{-1}$). The characteristic nanoscale thickness of vdW materials can be a drawback in terms of mechanical resistance, hence, it is vital to quantify their Young's moduli to consider their applicability in real-world devices (Table 2.2). For instance, tearing is a potential challenge when 2D materials are transferred on substrates, having to be avoided with stiff vdW structures to perform accurately.

Table 2.2 Young's moduli of different 2D materials with anisotropic behaviour

	Graphene		MoS_2		α-In_2Se_3		h-BN		References
	Z-Z	A-C	Z-Z	A-C	[100]	[001]	Z-Z	A-C	
E (TPa)	1.13	1.05	0.21	0.22	0.11	0.04	0.78	0.77	[37, 45, 46, 49]

In the case of graphene, multiple investigations have proved its remarkably high Young's modulus, positioning it as the second strongest material in the world, only behind carbyne, a one-dimensional (1D) carbon allotrope [33]. This quality gives graphene outstanding potential for flexible electronic and mechatronic applications. In the literature, the reported Young's modulus values range between 900 and 1100 GPa, thus, the Young's modulus of 1LG is commonly considered 1 TPa [34–36]. These quantities were obtained by a combination of experimental and modelling techniques like in situ tensile testing, peak force atomic force microscopy (AFM) imaging, finite element analysis (FEA), molecular dynamics (MD) simulations or elastic model fitting. Interestingly, the majority of the mechanical studies of graphene assume that the material is isotropic in the basal plane (its properties are the same for the longitudinal and transversal directions). However, given graphene's hexagonal lattice, the covalent bonds in the longitudinal trajectory follow a zigzag (Z-Z) path, whereas the bonds in the transversal orientation arrange in an armchair (A-C) disposition (see Fig. 2.1), breaking the sixfold symmetry for the physical properties to become intraplane anisotropic in finite vdW layers. This results in different values depending on the measurement's directionality. Ni et al. investigated this phenomenon via MD simulations and obtained Young's moduli of 1.13 and 1.05 TPa for the longitudinal and transversal directions, respectively [37]. Given the similarity in the in-plane covalent bindings, the values do not drastically differ, but, when considering the cross-plane direction, the weaker vdW bonds play their role by decreasing the measured quantities and creating a major anisotropy in the majority of the cases. Reports show that the cross-plane properties of 2D materials depend mainly on the number of layers and the type of interaction (lattice matching) between them and with the substrate [38, 39]. The reported values of bilayer graphene's (2LG) cross-plane Young's modulus range from 22.40 to 82.70 GPa depending on the density functional theory (DFT) functional used in the calculations [39].

Remarkably, the majority of TMDs share similar Young's modulus coefficients, given that their crystallinity is maintained along with their different compositions, keeping the covalent bonds as the main influencers to their notable mechanical behaviour. Various research groups have published similar Young's modulus results for monolayer MoS_2 of around 270 GPa [40, 41] (comparable to steel with ~ 200 GPa). These results were obtained by AFM nanoindentation measuring the deflection between the probe and the sample and considering the material as mechanically in-plane isotropic. Castellanos et al. studied the increase of Young's modulus in MoS_2 along with the thickness via AFM deflection and attributed this behaviour to the increase in bending rigidity due to the accumulation of consecutive monolayers [42, 43]. Other TMDs have also been the subject of mechanical studies. Deng et al. calculated (using first principles simulations) the in-plane Young's moduli of the four most recognised TMDs, being 217.60 GPa for MoS_2, 171.40 GPa for $MoSe_2$, 218.90 GPa for WS_2 and 201.50 GPa for WSe_2 [44]. Furthermore, Li et al. proved a certain Young's modulus anisotropy for different TMDs, being 222.75, 178.78, 244.18 and 196.81 GPa for the A-C direction and 219.46, 175.97, 240.99 and 194.13 GPa for the Z-Z direction for MoS_2, $MoSe_2$, WS_2 and WSe_2, respectively [45]. Correspondingly, the cross-plane Young's modulus value for MoS_2 is 100 GPa [41].

InSe is being used in several applications that require elastic semiconductor materials keeping a stiff Young's modulus value, hence, its mechanical properties are also an interesting subject of study. Raranskii et al. were the first research group to extract mechanical features from this material. They obtained highly anisotropic values for the Young's modulus of α-In_2Se_3, among other mechanical parameters. Using a combination of experimental tensor values and theoretical Voigt-Reuss-Hill (VRH) average, they determined Young's modulus values of 105.48 and 36.47 GPa for the [100] and [001] directions, respectively [46].

Being among the stiffest insulators in the world, h-BN also shows remarkable mechanical behaviour. Falin et al. investigated the variations of Young's moduli in h-BN with the addition of up to 9 monolayers using AFM nanoindentation (Fig. 2.5). They found that h-BN does not present a high variation in Young's moduli with the thickness increase due to the higher sliding interlayer energies that prevent the basal planes from sliding, resulting in increased cross-plane stiffness [47]. By contrast, Mirnezhad et al. proved a decreasing general trend of Young's modulus together with thickness in h-BN, taking into account the change in the stacked monolayers' order and using ab initio calculations [48]. Both of the previous references present similar Young's modulus values of 865 $\pm$ 73 and 829 GPa. Furthermore, Wu et al. used DFT to calculate the Young's modulus anisotropy along the two in-plane directions, showing a negligible variation of 780 $\pm$ 20 GPa for Z-Z and 773 $\pm$ 40 GPa for A-C directions, respectively [49].

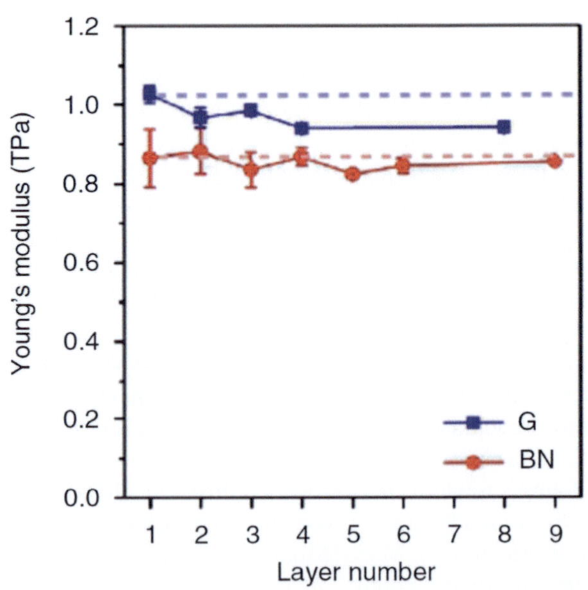

Fig. 2.5 Young's moduli versus layer number of graphene and h-BN. Adapted from Falin et al. [47], licensed under CC BY 4.0 [47]

Tensile Strength

Another essential characteristic of 2D nanomechanical systems is the ultimate tensile strength, also known as intrinsic strength (τ_T), with the same SI units as the Young's modulus. This property measures the maximum stress that can be applied to a material before it starts fracturing. In conjunction with the Young's modulus, this property is useful to determine the resilience of a measured 2D material since it is common to subject them to loads which can cause permanent deformation or fractures due to their low thickness and defect density (Table 2.3).

Tensile strength on graphene has also been calculated in many publications, with the majority of them coinciding at around 130 GPa. Lee et al. used AFM nanoindentation to measure this property, obtaining values of 130 $\pm$ 10 GPa for a 0.34 nm thick graphene layer [50]. An interesting dependence of the tensile strength with the anisotropy of graphene was studied via MD and orthogonal thigh-binding method simulations by Zhao et al., yielding values of 102 and 129 GPa for the A-C and Z-Z directions, respectively, which are in good agreement with the experimental values [51]. Furthermore, Wong et al. investigated the temperature-induced variation in the tensile strength of free-form 1LG submerged in water via MD simulations, showing a decay of the strength as the temperature increased (Fig. 2.6) [52]. Hossain et al. proved the Z-Z/A-C anisotropy in the intrinsic strength of graphene using computational simulations, having a 10% difference in the strength values and being the Z-Z direction the strongest one [53].

TMDs are also expected to achieve high tensile strength values given their notable Young's modulus. Bertolazzi et al. conducted AFM nanoindentation to obtain the tensile strength of monolayer MoS_2, which resulted to be a 10% of the obtained Young's modulus value, yielding a tensile strength range of 16–30 GPa [40]. To quantify the anisotropy of this property in different TMDs, Li et al. used first-principles density functional calculations. For MoS_2, the tensile strength was found to be 27.35 GPa in the A-C direction and 16.90 GPa in the Z-Z direction, whereas, for $MoSe_2$, WS_2 and WSe_2, the values were 22.68, 29.96 and 24.70 GPa for their A-C directions and 12.86, 19.91 and 15.05 GPa for their Z-Z directions, respectively, proving considerable anisotropy in these materials [45].

h-BN does not fall behind in showing notable mechanical tensile strength. Falin et al. provided results for the tensile strength versus thickness dependence of h-BN, showing values of 70 $\pm$ 5 GPa, concluding that the tensile strength of h-BN does not vary with thickness as much as graphene does, in agreement with the Young's modulus [47]. Wu et al. found a moderate in-plane anisotropy for the intrinsic strength in monolayer h-BN, being 102 and 88 GPa for the Z-Z and A-C directions [49].

Table 2.3 Tensile strength of different 2D materials with in-plane anisotropic behaviour

	Graphene		MoS_2		h-BN		References
	Z-Z	A-C	Z-Z	A-C	Z-Z	A-C	
τ_T (GPa)	129	102	12.86	22.68	102	88	[45, 49, 51]

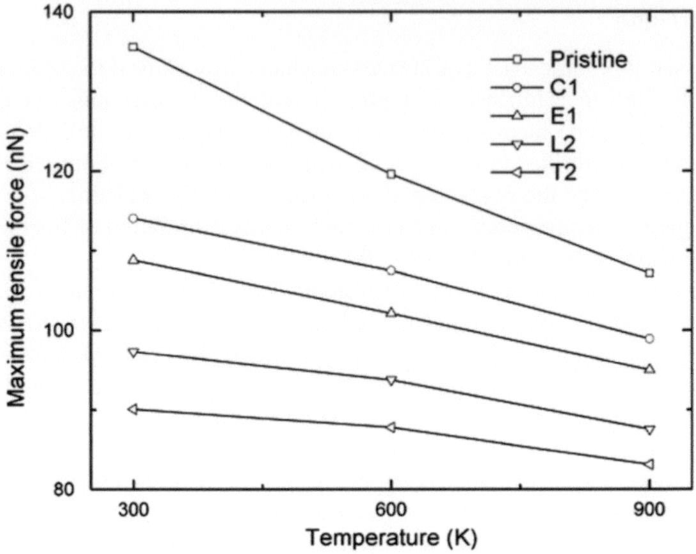

Fig. 2.6 Tensile strength in graphene against applied temperature. Reproduced with permission from Ref. [52], Copyright © 2012 Elsevier [52]

2.2.3 *Electrical Transport*

Electrical Conductivity

The ability of any given material to allow a charge carrier current (positive or negative) to flow is called electrical conductivity (σ), with SI units of S m^{-1}, defining the easiness of transmitting electricity. This property is an intrinsic parameter in each element at the bulk scale, meaning that a material will have constant electrical conductivity even if its size and morphology vary. Interestingly, when the dimension of the material changes from bulk to 2D, novel phenomena occur, causing a variation in the electrical conductivity (Table 2.4). In these low dimensions, the charge carrier density and mobility dominate the electrical transport within the material, taking the leading role over its electrical conductivity [9, 54].

Graphene is one of the best electrical conductors known to date. This is due to its characteristic sp^2 hybridisation (Fig. 2.7), which leaves a delocalised π-orbital in the cross-plane direction and results in a free movement of the electrons in these orbitals in both faces of the basal plane [8]. Additionally, the zero bandgap in graphene

Table 2.4 Electrical conductivity of different 2D materials

	Graphene	MoS_2	α-In_2Se_3	h-BN	References
σ (S m^{-1})	2.40×10^5	4.50×10^4	6.89×10^3	10^{-11}	[57, 62, 64, 65]

results in a linear dispersion relation between the energy and momentum in its electronic spectrum, crossing at a single point in the Brillouin zone known as Dirac point. Here, electrons travel ballistically with minimal scattering, resembling massless Dirac fermions near the Fermi level, which, combined with the observation of quantum Hall effect, increases the mobility and its electrical conductivity. Despite a broad investigation in this aspect, the maximum value of electrical conductivity of pristine graphene has not yet been accorded due to the complexity of the measurement [55]. The most common research methodology requires the creation of graphene composites with different materials (usually metals or polymers) to enhance their electrical properties [55, 56]. Nonetheless, an approach for the study of the electrical conductivity of a suspended graphene monolayer was performed by Wang et al. [57]. They used an experimental T-type sensor methodology to measure the electrical conductivity with respect to the temperature, yielding a value of 2.40×10^5 S m^{-1} at 33 °C. Furthermore, Murata et al. investigated the electrical properties of FLG via Hall effect measurements in order to relate the thickness of the graphene films with the electrical conductivity. They found the highest electrical conductivity for 50 nm thick graphite of 2.70×10^5 S m^{-1} [58]

Besides its high electrical conductivity, graphene does not possess a suitable on/off ratio due to its null bandgap. In this regard, multiple TMDs have become a useful substitution tool for semiconductor applications since they have tunable bandgaps that allow for modulated electrical conductivities via doping or strain [59, 60]. Chen et al. studied the thermoelectric properties of $MoSe_2$, trying to increase the power factor by improving the Seebeck coefficient (S) and the electrical conductivity. They used a wide range of experimental characterisation techniques to measure the electrical conductivity with respect to the number of $MoSe_2$ layers, finding a decrease in the values from 1.50×10^5 to 460 S m^{-1} as the thickness increased from 6 to 2.70×10^3 nm [61]. Similarly, Hippalgaonkar et al. experimentally measured the dependence of electrical conductivity versus thickness of MoS_2 down to the monolayer

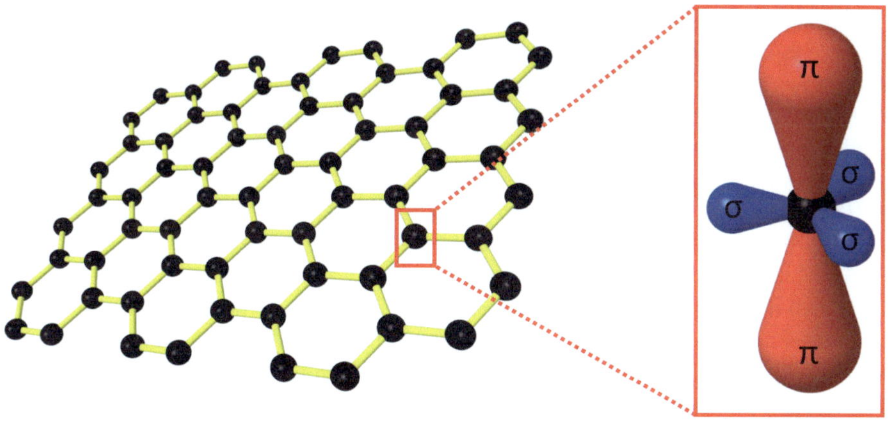

Fig. 2.7 Bonding orbitals of carbon atoms in graphene

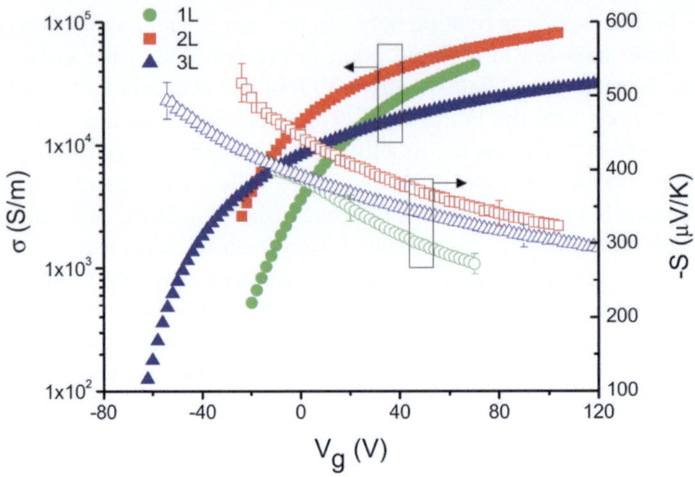

Fig. 2.8 Electrical conductivity and Seebeck coefficient versus gate voltage and film thickness of MoS_2. Reproduced with permission from Ref. [62], Copyright © 2017 American Physical Society [62]

scale, simultaneously varying the gate voltage (V_g) and the temperature, resulting in different Seebeck coefficient values (Fig. 2.8). They showed an increasing conductivity together with thickness, temperature and charge carrier concentration for negative gate voltages [62]. Analogous results were found by Schmidt et al. examining the conductivity of monolayer WSe_2 and MoS_2 with respect to the temperature and gate voltage change. They found a reversible behaviour for these two materials, being WSe_2 an insulator at positive gate biases and a conductor at negative ones, whereas the behaviour was reversed for MoS_2 [63].

InSe shows interesting electrical behaviour when the polymorphic conductivities are compared. Kim et al. studied the electrical conductivity of α-In_2Se_3 single crystals via Hall effect measurement, comparing it to the change in temperature, yielding a range of 9.45×10^3 to 6.89×10^3 S m^{-1} at 15 and 300 K, respectively, cataloguing the single crystal as a n-type degenerate semiconductor with conductivity similar to metallic materials due to its high doping, which positions the Fermi level in the conduction band [64]. Also, a study done by Emziane et al. showed the variability in electrical conductivity for α, β and γ phases of In_2Se_3 thin films, probing values of 100, 1 and 0.10 S m^{-1}, respectively [65].

h-BN is known to possess a wide direct bandgap that positions the material as an electrical insulator [66]. Thus, its electrical behaviour is treated in the literature from the resistivity point of view instead of its inverse, the electrical conductivity. Steinborn et al. measured via resistance meter the change in the electrical resistivity in different microstructures of h-BN, obtaining a range of 10^{11} to 10^{13} Ωm while varying the type and quantity of impurity defects [67]. Uddin et al. focused their study on the improvement of h-BN electrical conductivity by doping it with carbon atoms. They managed to reduce the electrical resistivity of the alloy at room temperature (RT) from 109 to 0.10 Ωm by an approximate 20% increase in the doping concentration. Furthermore, they also measured the variation of resistivity whilst the material's temperature changed, finding a decrease from 0.16 to 0.03 Ωm with a temperature increase from 650 to 820 K [68]. Wang et al. measured the effect of doping h-BN with beryllium, magnesium and zinc on the electrical resistivity, yielding values of 430, 0.001 and 0.12 Ωm, respectively [68].

Charge Carrier Mobility and Density

The current distribution in low dimensional materials is ultimately defined by two parameters, the carrier mobility (μ) and the carrier density (n), with SI units of m^2 V^{-1} s^{-1} and m^{-3}, respectively (Table 2.5). In the majority of materials, the transfer of electrical charges is conducted by elementary charged particles, i.e., electrons and/or holes, with negative and positive charges, respectively. These particles are quantifiable by their ability to be transferred amongst the material via an electric field (charge mobility) and by their concentration (charge density). The electrical conductivity depending on the charge mobility and density can be expressed as [69, 70]:

$$\sigma = qn\mu \tag{2.1}$$

where q is the elementary electric charge (1.60×10^{-19} C).

The electronic transport in graphene has been widely studied over the recent years, both theoretically and experimentally, with remarkable results showing an exceptionally high carrier mobility even at high charge densities, a low reactivity towards chemical doping and a ballistic transport at RT and submicron scale [2]. Novoselov et al. used field-effect and magnetoresistance measurements to report charge mobilities of up to 1 m^2 V^{-1} s^{-1} for FLG at a corresponding in-plane carrier density of 5×10^{16} m^{-2}, showing also negligible temperature versus mobility dependence [4, 71]. Although the intrinsic mobility of suspended 1LG was found to be 20 m^2 V^{-1} s^{-1} with a charge density of 2×10^{15} m^{-2}, the more realistic combination of the graphene layers with various supporting substrates has been extensively studied with the goal of enhancing the 2D material's mobilities for practical application in devices, achieving values of 54 m^2 V^{-1} s^{-1} at a concentration of 10^{15} m^{-2} and 100 K by covering 1LG with chromium oxychloride (CrOCl) and h-BN insulator layers [72]. Graphene also shows in-plane anisotropy when it comes to the mobility of its carriers. Armakavicius et al. calculated the mobility variation of the two in-plane

Table 2.5 Charge carrier mobility and corresponding density of different 2D materials

	Graphene	MoS_2	γ-InSe	h-BN	References
μ (m^2 V^{-1} s^{-1})	20	1.79×10^{-3}	0.10	1.85×10^{-3}	[72, 75, 77, 80]
n (m^{-2})	2×10^{15}	10^{16}	7.90×10^{16}	0.80×10^{11}	

directions of 1LG via terahertz–optical Hall effect measurements at a charge density of ~ 3×10^{16} m^{-2}, finding values of 0.11 and 0.10 m^2 V^{-1} s^{-1} for the x and y directions, respectively [73].

Due to the suitability of monolayer TMDs for semiconductor applications regarding their direct bandgap, several studies aim to increase their charge carrier mobility for improved electrical transport. Cheng et al. reported RT longitudinal acoustic and optical phonon limited electron mobility values of several TMDs (MoS_2, $MoSe_2$, WS_2 and WSe_2 amongst them), calculated via the density functional perturbation theory. $MoSe_2$ was reported to have the lowest mobility of the four mentioned TMDs, with 0.01 m^2 V^{-1} s^{-1}, followed by WSe_2 with 0.02 m^2 V^{-1} s^{-1}, MoS_2 with 0.04 m^2 V^{-1} s^{-1} and WS_2 with 0.08 m^2 V^{-1} s^{-1} [74]. The intrinsic mobility of MoS_2 was experimentally studied via RT Kelvin probe force microscopy (KPFM) by Dagan et al. considering a charge density of 10^{24} m^{-3} [75]. With this methodology, they were able to eliminate the mobility hindrance effects of the electrode/sample contact quality and the gate voltage screening ability, obtaining accurate charge mobility values of up to 1.79×10^{-3} m^2 V^{-1} s^{-1} for monolayer MoS_2. Furthermore, Rawat et al. studied the anisotropic effects on the charge mobility of electrons and holes in RT MoS_2, stating values of up to 11.07×10^{-3} and 28.52×10^{-3} m^2 V^{-1} s^{-1} in the Z-Z direction and 10.78×10^{-3} and 30.86×10^{-3} m^2 V^{-1} s^{-1} in the A-C direction for electrons and holes, respectively [76].

An alternative to TMDs in semiconductor applications that require higher charge mobility can be InSe, which, in its few-layer γ form, has proven to have values comparable to WS_2 at RT (~ 0.10 m^2 V^{-1} s^{-1}) and up to 1.27 m^2 V^{-1} s^{-1} at 25 K [77]. However, most of the related carrier mobility studies have been conducted using theoretical methods. Nonetheless, Li et al. developed an InSe field-effect transistor with an additional indium layer that reduced the gold/InSe contact barrier and induced a surface electron-doping to enhance the mobility achieving unprecedented values of up to 0.37 m^2 V^{-1} s^{-1} at RT and n of 1.30×10^{16} m^{-2} [78].

Due to the low electrical conductivity of h-BN, several research groups are trying to increase its naturally low value of charge mobility via doping, temperature optimisation or combination with other materials. The phonon-limited mobility of electrons and holes on h-BN was calculated via first-principles studies at RT by Khatami et al., resulting in 11.80×10^{-3} m^2 V^{-1} s^{-1} for electrons and 44.40×10^{-3} m^2 V^{-1} s^{-1} for holes [79]. Additionally, Doan et al. measured the concentration and hole mobility of h-BN bulk epilayers at high temperatures, finding a maximum mobility value at 700 K and 0.80×10^{19} m^{-3} of 1.85×10^{-3} m^2 V^{-1} s^{-1} [80].

2.2.4 *Heat Transport and Thermal Properties*

Thermal Conductivity

Due to the high industrial interest in energy conservation, thermal conductivity (κ) has been a major subject of study, especially in 2D materials (Table 2.6). This parameter quantifies the capability of heat flux transmission in a material in units of W m^{-1} K^{-1}. In solid-state materials, thermal conductivity is determined by both electronic and phononic heat transport. Electrons are mainly known for their charge transport ability, however, they can also be heat carriers performing thermal transport within the material, depending on their mean free path (λ). Phonons are essentially lattice vibrations that also contribute to heat transfer and can be subsequentially defined by two main factors, the density of states (DOS) and the states' energy. The DOS depicts the different phonon vibrational modes in the material, hence, with a higher DOS, the number of vibrational modes and phonons contributing to the heat transport increases. Similarly, the energy of these states becomes beneficial for heat transfer if it is high. Generally, the thermal conductivity in metallic materials is electronically dominated, given the higher density of valence electrons capable of conducting heat and their greater mean free path. Nevertheless, the movement of electrons in insulating materials is highly restricted, hence, the phonon contribution increases due to the relevant DOS and vibration energy.

Graphene's thermal conductivity has been extensively measured due to its superior heat transport behaviour, extrapolating to applications such as energy harvesting. In order to study graphene's intrinsic thermal behaviour, Lee et al. conducted Raman scattering spectroscopy of suspended pristine 1LG, comparing the thermal conductivity with the applied temperature. They used heat diffusion equations to obtain thermal conductivity values of 1800 and 710 W m^{-1} K^{-1} at 325 and 500 K, respectively [81]. Balandin et al. performed similar procedures to find different thermal conductivity values of $(4.84 \pm 0.44) \times 10^3$ to $(5.30 \pm 0.48) \times 10^3$ W m^{-1} K^{-1} at RT changing the laser excitation power (source of the variability in the thermal conductivity values) [82]. Lindsay et al. reviewed the variation of thermal conductivity on 1LG and FLG with a size and thickness dependence via Boltzmann transport equation calculations (Fig. 2.9). They showed an increase of the thermal conductivity with the reduced number of layers and larger sized materials [83]. Cui et al. found in-plane quasi-isotropic behaviour on pristine graphene's thermal conductivity at RT, quantified as 180.64 and 178.90 W m^{-1} K^{-1} for the A-C and Z-Z directions, respectively. Interestingly, when the layers were subjected to linear strain, the heat transport became anisotropic, being the Z-Z direction better thermal conductor than the A-C one [84]. Conversely, the cross-plane thermal conductivity value of FLG was found to be much lower (~ 6 W m^{-1} K^{-1}) [85].

In contrast to the high in-plane thermal conductivity of graphene, which makes it unsuitable for applications like thermoelectricity, TMDs generally possess lower values together with suitable electrical behaviour, making them a valuable asset for

Table 2.6 Thermal conductivity of different 2D materials with anisotropic behaviour

	Graphene		MoS_2		γ-InSe		h-BN		References
	xy	z	xy	z	xy	z	xy	z	
κ (W m^{-1} K^{-1})	1800	6	82.20	3	8.50	0.76	2000	0.20	[22, 81, 82, 87, 90, 92, 95]

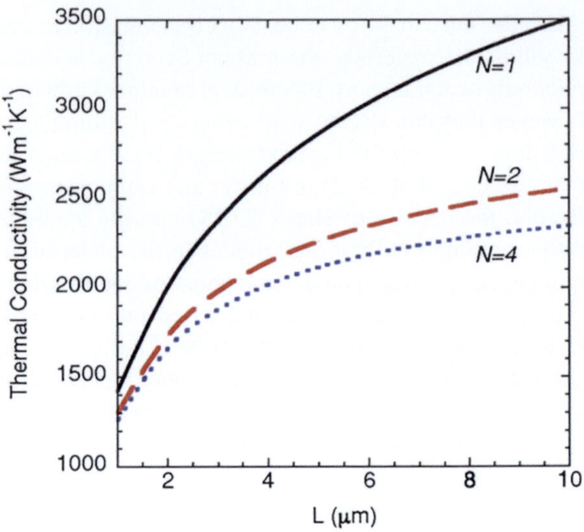

Fig. 2.9 Dependence of the thermal conductivity versus thickness and length of graphene layers. Reproduced with permission from Ref. [83], Copyright © 2011 American Physical Society [83]

this type of application. The thermal conductivity of few-layer MoS_2 was measured by Sahoo et al., extracting the temperature coefficients in the in-plane and cross-plane directions and performing laser-power-dependent Raman scattering, yielding a final RT in-plane thermal conductivity of 52 W m^{-1} K^{-1} [86]. Zhang et al. studied the thermal conductivity of MoS_2, $MoSe_2$, WS_2 and WSe_2 among others [87]. They used the Boltzmann transport equation combined with first-principle calculations to extract values of 82.20, 46.20, 121.20 and 72.70 W m^{-1} K^{-1} for MoS_2, $MoSe_2$, WS_2 and WSe_2, respectively, at RT. Elsewhere, Pallecchi et al. reviewed the thermal conductivity of TMDs and compared them to variables such as doping, off-stoichiometry, and sample morphology [88]. They showed an experimental thermal conductivity range of 35–55 W m^{-1} K^{-1} for MoS_2 and a record low thermal conductivity of 0.05 W m^{-1} K^{-1} for WSe_2 along the cross-plane direction. However, various studies limit the cross-plane thermal conductivity of MoS_2 in the range of 2–3 W m^{-1} K^{-1} [89, 90].

As a post-TMD, InSe shares comparable properties to those of MoS_2, with thermal transport not being an exception. In this sense, InSe is even capable of better heat management maintaining proper electrical behaviour, resulting also in its adequate applicability in thermoelectric technologies. Zhou et al. made use of micro-Raman spectroscopy and FEA to determine the thermal conductivity of β-In_2Se_3 crystal layers depending on their thickness [91]. They obtained in-plane values of 60 and 4 W m^{-1} K^{-1} for 35 and 5 nm thick layers, respectively, proving the importance of phonon surface scattering and thermal conductivity relation. Additionally, γ-InSe presents anisotropy in its thermal conductivity since the in-plane and cross-plane values are 10.42 and 1.74 W m^{-1} K^{-1}, respectively [19], although a more recent publication achieves diminished κ of 8.50 ± 2 in-plane and 0.76 ± 0.15 W m^{-1} K^{-1} cross-plane [92]. Furthermore, Hsin et al. discovered a notable thermal conductivity anisotropy for nanobelts of α-In_2Se_3 through temperature-dependent measurements via a suspended pattern technique [93]. The thermal conductivity of the nanobelts increased from 3 to 3.50 W m^{-1} K^{-1} at a temperature range of 300–400 K with a tenfold difference in the direction due to the effect of vdW force scattering between the selenium/selenium interfaces.

The wide bandgap of h-BN makes it an electrical insulator, but it does not impair its heat properties, in fact, its thermal conductivity can be comparable to graphene. Lin et al. reviewed the properties of h-BN in different structural configurations [22]. They stated that the theoretical thermal conductivity of h-BN ranges between 300 and 2000 W m^{-1} K^{-1}, differentiating from graphene with softer phonon modes and different masses between boron and nitrogen atoms. Wang et al. measured the RT thermal conductivity of h-BN using suspended prepatterned microstructures. They quantified it in the range of 460–625 W m^{-1} K^{-1} and compared it to different thicknesses, finding that the bilayer exceeded the bulk κ [94]. Jaffe et al. experimentally measured the RT cross-plane thermal conductivity of 7 nm thick h-BN to be 0.20 ± 0.06 W m^{-1} K^{-1} [95].

Interfacial Thermal Insulance

2D materials have to interact with other types of structures, either other 2D layers or bulk substrates, forming a transition between the blocks known as junction or interface. Said interface depicts the mismatch between the surface properties of the interacting materials. Hence, in nanoscale structures where most of the composing atoms are in the surface (high surface/bulk ratio), the role of these interfaces on the material's properties becomes prominent. Interfacial thermal insulance (r_{int}) is an intrinsic property that defines the resistance per unit area to the thermal flow at this interface generated by differences in the electronic and lattice vibrational structure of the involved materials, and its SI units are km^2 W^{-1}. These variations originate in the available energy states of each material, which, depending on their quantity, will prompt a higher or lower transmission probability of thermal energy carriers (phonons or electrons). Depending on the type of material, the energy transmission will be dominated by electrons or lattice vibrations, which will scatter at the interface, creating a thermal resistance and a temperature jump.

Given that this parameter only exists with an interaction between two systems, Table 2.7 depicts the literature values of the stated 2D materials assembled with themselves, other 2D layers or bulk substrates forming homo- or heterojunction structures, respectively.

Table 2.7 Interfacial thermal insulances of different 2D materials' junctions

Junction	$r_{int} \times 10^{-8}$ ($km^2 W^{-1}$)	Methodology	References
Graphene/graphene	5.20	Thermoreflectance	[96]
Graphene/MoS_2	10	Thermoreflectance	[96]
Graphene/h-BN	1.84–1.99	Raman spectroscopy	[97]
Graphene/SiO_2	3.92	Thermoreflectance	[98]
Graphene/silicon	14.29	Thermoreflectance	[99]
MoS_2/MoS_2	135	Optothermal Raman	[100]
MoS_2/h-BN	5.75–6.02	Raman spectroscopy	[97]
MoS_2/SiO_2	5.60–10	Raman thermometry	[101]
MoS_2/gold	196.10–270.30	Optothermal Raman	[100]
h-BN/SiO_2	1.60	3ω-method	[102]

2.3 Preparation of Two-Dimensional Materials and Nanodevices

2.3.1 Exfoliation, Transfer and Stacking

The production of 2D materials can be systematically divided into two main approaches, called top-down and bottom-up. In the bottom-up approach, 2D layers are formed by assembling a set of smaller elements, e.g., grouping atoms or molecules. Fabricating with this methodology allows to have great control, tunability and precision over the final structure to the detriment of time and budget optimisation. This approach is used in chemical vapour deposition (CVD) to epitaxially assemble layers of 2D materials on top of a substrate or by the sublimation of silicon atoms in a silicon carbide (SiC) layer, leading to a graphitisation of the remaining carbon atoms [103]. On the contrary, top-down methods are more time- and budget-efficient, but the quality of the resulting structure is considerably reduced. In this approach, the initial bulk material is decreased in dimensions until the nano-range is reached.

In exfoliation, a few layers are separated from the bulk crystal breaking the interlayer vdW bindings and deposited onto a substrate. There is a wide catalogue of exotic exfoliation techniques (electrostatic force assisted exfoliation [104], anodic bonding [105], electron-irradiation-induced charging [106], etc.), but the most common approaches are chemical or liquid-phase exfoliation (LPE) and mechanical exfoliation or cleavage.

As the original method [4], mechanical exfoliation has been the most replicated process for 2D materials' production, resulting in small flakes with a low yield. This method consists of the application and peeling of an adhesive tape from the surface of a bulk crystal, detaching some 2D flakes (Fig. 2.10a–c). The process can be repeated by folding the tape to reduce the thickness.

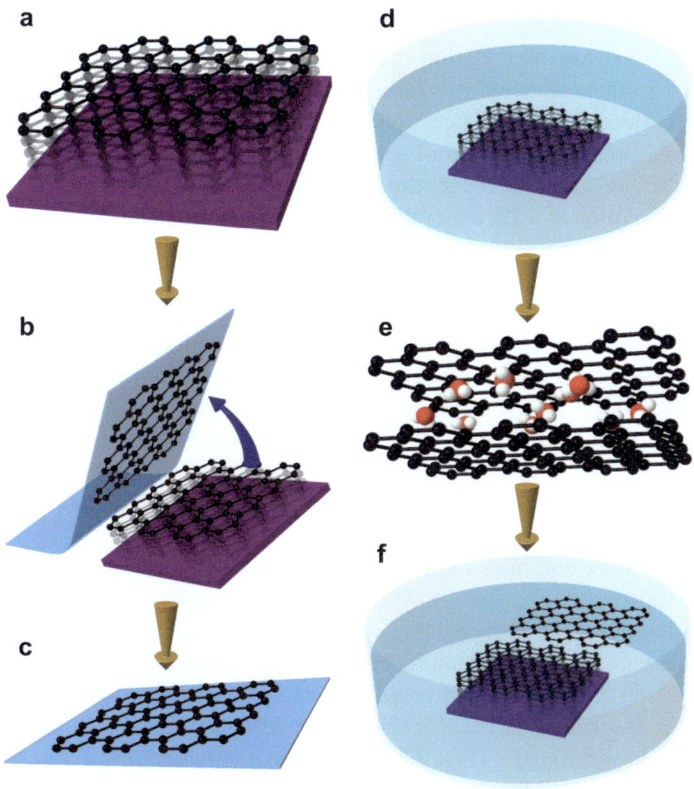

Fig. 2.10 2D material exfoliation. Mechanical exfoliation: **a** stacked 2D layers on a substrate, **b** monolayer exfoliation with tape and **c** monolayer on the tape. Liquid-phase exfoliation: **d** Stacked 2D layers on a substrate in a solvent, **e** solvent molecules intercalating between the layers and **f** monolayer floating on the liquid's surface

LPE generally yields smaller flakes than mechanical exfoliation, however, it is still widely used in many research groups. The process begins with the dispersion of bulk crystals in a solvent with similar surface energy (to match the vdW interlayer binding energy and prevent layer re-aggregation). Ultrasonic or shearing energy is then supplied to aid the intercalation of the solvent between the 2D layers and to form high surface and kinetic energy cavitation bubbles within the liquid which fragment the vdW bonds, causing the removal of nanosheets from the bulk crystals. Such thin structures are ultimately separated and collected by a centrifugation step from the unexfoliated layers (Fig. 2.10d–f) [107].

Electrochemical exfoliation is another relevant method which uses an external power source to separate the layered materials submerged in electrolyte through direct or bipolar configurations. In the direct setup, a bias applied to the material causes the electrolyte's ionic species insertion and gas bubble formation, leading to layer detachment. Bipolar exfoliation involves placing the material between two feeder electrodes, creating a uniform electric field that induces polarity on the layered material and drives ion intercalation as well as redox reactions to promote exfoliation. Similarly, hydrothermal exfoliation employs high temperature and pressure ions that intercalate between the nanosheets and electrostatically repulse to promote their separation. Although a variation of such method where intralayers are chemically etched with acidic solutions before the ion intercalation can also be employed [108].

After the isolation of a 2D material from its bulk crystal, the process of deposition of this ultrathin layer onto a suitable substrate is known as transfer, and it varies depending on the previously used isolation method. Prior to any transfer procedure, the target substrate must be clean and surface-activated to achieve an effective deposition yield. Depending on the preliminary level of contamination on the substrate's surface, a chemical treatment sonication with acetone, isopropanol (IPA), ethanol (EtOH) and/or deionised water (DIW) is the first step, followed by a neutral gas drying of the chemicals (via an argon or nitrogen gun). The last step is the activation of the surface, which is done by a plasma etching treatment, disrupting the oxygenated terminations, creating active bonds and eliminating most of the organic pollutants remaining on the substrate [109].

One of the most common transfer methods to the active substrate is the direct exfoliation transfer (Fig. 2.11a–c). This procedure is performed after the 2D material mechanical cleavage with an adhesive tape, which is then applied on the substrate together with constant pressure and/or heat to improve the 2D deposition yield [109]. Another transfer methodology involves the use of a liquid matrix to aid with the separation of the layered structure from a growth substrate, splitting into wet or dry transfers. In wet transfer (Fig. 2.11d–h), poly(methyl methacrylate) (PMMA) is generally used to shield the 2D material and facilitate the transfer. By submerging the PMMA/2D material/growth substrate ensemble into a solvent, the PMMA/2D material layer detaches from the growth substrate and can be transferred to water to "fish" it with the target substrate [110]. Leong et al. employed Paraffin wax in exchange for PMMA to avoid the polymer's contamination, the creation of wrinkles and other adverse effects [111]. Conversely, in dry transfer, the 2D material is never in contact with the solvent, given that it only detaches the PMMA/2D material structure by dissolving the releasing layer between the PMMA and the growth substrate (Fig. 2.11i–m) [110].

Given the vdW vertical arrangement nature of these materials, researchers have started exploring the possibilities of manually stacking two or more 2D layers as building blocks. This new postproduction branch creates heterostructures if the stacked materials are different, e.g., graphene on h-BN, or homostructures if the composition is identical, e.g., MoS_2 on MoS_2. These structures generate novel and exciting properties beyond the monolayer and a higher number of degrees of freedom for device engineering. Due to different commensurate states between the building blocks, the final vertical ensemble is arranged with weak vdW interlayer attraction. Regardless of the interfacial lattice matching constraints, vdW heterostructures form neat interfaces, improving the tunnelling between layers even when interlayer defects are present. Nevertheless, the key attribute of these structures is their tunability,

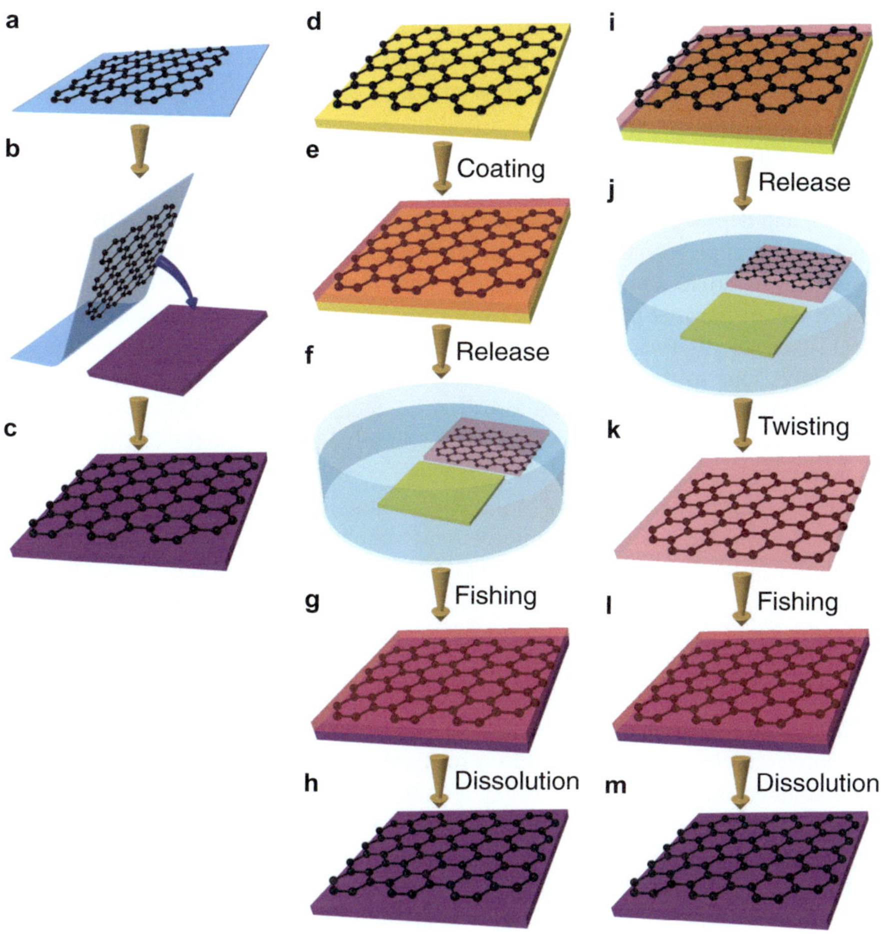

Fig. 2.11 2D material transfer. Direct exfoliation transfer: **a** pre-exfoliated 2D material on the tape, **b** monolayer and tape attaching to the substrate and **c** transferred monolayer after the tape is pressed and peeled off. Wet transfer: **d** 2D material on a growth substrate, **e** coated PMMA on the 2D material/growth substrate ensemble, **f** liquid-aided released PMMA/2D material, **g** deposited PMMA/2D material layer on target substrate after its solvent "fishing" and **h** 2D layer transferred on the target substrate after the PMMA dissolution. Dry transfer: **i** 2D material on a PMMA/ release layer/growth substrate ensemble, **j** released 2D material/PMMA layer after the solvent-aided release layer's dissolution, **k** 2D material/PMMA layer before its transfer to a neutral post-solvent, **l** deposited PMMA/2D material layer on target substrate after its post-solvent "fishing" and **m** 2D layer/target substrate after the PMMA dissolution

leading to "on-demand" properties depending mostly on the stacking sequence, the relative interlayer rotation, the interlayer spacing and the layer composition [112].

The fabrication of these stacked systems is nonetheless intricate. Analogous to exfoliation, numerous methodologies have been proposed, e.g., electrostatic assembly or direct growth assembly [112], but the prevalent processes are still encompassed by a mechanical transfer due to its simplicity compared to the other mentioned procedures. Such mechanical stackings are based on precise pick-up and drop-down steps, assembling the system with a bottom-up approach. The method primarily consists of the top-down exfoliation of 2D layers from a bulk crystal with a subsequent transfer to a target substrate, repeating this step with a different or identical material and depositing the second layer on top of the previous one until the desired structure is completed. The carrier method (Fig. 2.12a–d) is employed to fabricate homo- and heterostructures, commencing with the exfoliation of the vdW material via PMMA or thermal release tape from the growth substrate to deposit it onto the target substrate, afterwards, the PMMA can be dissolved, or the thermal release tape sublimated. Due to the lack of mechanical stability of the PMMA and the thermal release tape, an additional stiffer substrate is occasionally needed, contacting the polymer via a sacrificial layer, which can be dissolved to release the PMMA and the 2D material [112].

The low mechanical stability of the carrier transfer polymer hinders the epitaxy control. To avoid this drawback, microcontact stamping (Fig. 2.12e–h) makes use of transparent poly(dimethylsiloxane) (PDMS) as the transfer substrate, simultaneously eliminating the need for a polymeric chemical dissolution. This process allows for controlled manipulation in every step and only depends on the adhesion properties of the PDMS with the 2D material and the 2D material with the target substrate [112]. The process starts with the PDMS coating and rapid thermal-aided exfoliation of the 2D material from the growth substrate, occasionally adding a sacrificial polymer like poly(propylene carbonate) (PPC) to increase adhesion, and finishes with the thermal-aided transfer of the material, a continuous slow release of the PDMS and dissolution of the PPC (if used), leaving only the 2D layer on the target substrate or previously transferred 2D structures.

These are generally meticulous processes since the outcome highly depends on the lattice matching and the purity of the system. In this regard, the lattice matching can be understood as the compatibility between the building blocks, which encompasses the orientation of the layers, their lattice matching, the interlayer distance, etc. The purity represents the density of contamination present in the final structure between the layers, affecting the properties of the system. This contamination is typically generated due to the contact between the phase-changing polymeric sacrificial layers or dust particles and the 2D materials. Hence, the total efficiency of the stacking process is engineered by employing non-reactive materials or working in a pollutant-free environment and ultimately maximising the total compatibility of the building layers utilising precise transfer equipment like micromanipulator stations.

2.3.2 *Chemical Vapour Deposition*

The methodologies elucidated in the previous section can be a resourceful option for a rapid assembly of a layered sample, however, these structures present a series of

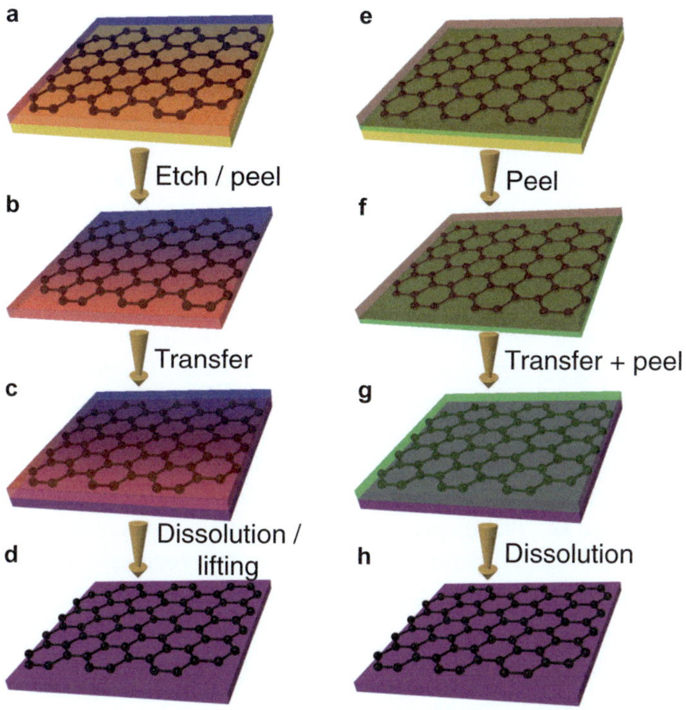

Fig. 2.12 vdW structures transfer. Carrier transfer: **a** PMMA or thermal release tape/2D material/growth substrate ensemble, **b** isolated PMMA or thermal release tape/2D material layer after etching of the substrate or peeling off from it, **c** transferred PMMA or thermal release tape/2D material layer on the target substrate (aiding with a robust substrate and a sacrificial layer is advisable) and **d** 2D material deposited on the target substrate after dissolving the PMMA or sublimating the thermal release tape. Microcontact stamping: **e** PDMS/PPC/2D material/growth substrate ensemble, **f** PDMS/PPC/2D material layer after peeling it from the growth substrate, **g** PPC/2D material layer on the target substrate after the slow PDMS detachment from the ensemble and **h** 2D material deposited on the target substrate after dissolving the PPC

drawbacks related to the quality and the dimension of the final product. In exfoliation, due to its intrinsic top-down nature, the material has to be separated from its original structure, disrupting the internal bindings and introducing impurities, which decreases its size and crystallinity and results in a poorly controlled process where most of the final layers have stochastic characteristics.

Chemical vapour deposition (CVD) emerges to offer an alternative technique capable of precisely controlling the parameters needed for 2D material growth. CVD makes use of vapour-phase precursors to deposit uniform and conformal thin films on pretreated substrates, therefore being classified as a bottom-up technique. During the process, features such as the film thickness, its composition and its overall structure can be meticulously monitored and optimised, generating materials with tailored properties [113].

Preliminary treatment of the target substrate must be performed to remove any contaminants and increase its epitaxy with the layered material. This usually commences by selecting the appropriate substrate depending on its catalytic binding with the deposited gas, followed by electrochemical polishing or decontamination processes like solvent cleaning or picking. Subsequentially, the substrate is subjected to an elevated temperature (> 900 °C) annealing in a stable atmosphere. Oxygen molecules can be introduced to burn any remaining carbon contaminants, and hydrogen can aid in activating the surface by reducing the oxygen terminations. The high temperature enhances the formation of single crystals in the substrate, which is beneficial for a better deposition of the material.

The main synthesis can then start with the decomposition of a volatile precursor inside a reaction chamber that contains the heated substrate. The precursor molecules generally induce a phase change into a gaseous state and are adsorbed onto the active substrate surface via a triggered chemical reaction, in this way, they bind not only with the substrate but also with the pre-deposited material by nucleation and coalescence, forming a solid thin film with controlled dimensions. In this deposition step, the reaction conditions, i.e., the temperature, pressure, molecule concentration, etc. are carefully controlled. Lastly, additional steps such as fast-cooling or annealing can be performed to modify the ultimate properties of the thin film [113].

2.3.3 *Optical Lithography Patterning*

The in-plane geometry of 2D structures has a crucial impact on their physical properties, generally defined by constrictions or boundary effects, and on their dimensional compatibility with specific applications or characterisation techniques. Shaping these 2D materials into specific lateral dimensions and applying a linkage between the nano- and macroscale, which can act both as input and output of signal, are the fabrication processes for 2D devices. Given the nanoscale dimensions of these structures, no tools can be directly manipulated by the researcher's hands to build such precise systems. Hence, new methods capable of high precision manufacturing are needed. In this regard, patterning has become the most adequate fabrication process. It consists of the rendering of specific shapes in a target material with the use of precisely controlled probes with a nanoscale diameter. In the present and following two sections, the most known pattering techniques of 2D materials are discussed.

Optical lithography or photolithography patterning is a process intended to create fine-structure patterns by using light as the lithographic probe. This process requires a prepatterned optical mask that shields the entire surface of the sample except for the areas that will be rendered, and the deposition of a sacrificial layer, known as photoresist, on top of the target material. This layer is composed of a light-sensitive material, meaning that its structure will change when it interacts with the beam photons that cross the voids of the optical mask pattern. Depending on how the exposed structure changes, i.e., by disrupting or creating interatomic bonds, the sacrificial layer can be respectively classified as positive or negative photoresist.

The sample is subsequentially subjected to chemical treatments, named developing steps, which remove either the disrupted areas of the positive resist or the unexposed areas of the negative resist, leaving specific regions of the target sample exposed for additional etching or material deposition processes [114].

The diffraction limit fundamentally constricts the obtainable photolithography resolution, which, according to the Rayleigh criterion, is proportional to the ratio between the wavelength of the exposure light and the numerical aperture of the imaging system [115]. Such lithographic diffraction limit is defined by the spreading of the light waves when they encounter small apertures or edges, following the Huygens' principle. This diffraction phenomenon causes the light to spatially expand and blur losing intensity, hindering its sharp focus on small features.

For this reason, the resolution limit in conventional optical lithography (with wavelengths from 400 to 248 nm) is $\geq$ 200 nm. To achieve smaller resolvable feature sizes, the exposure wavelength must be decreased, and the numerical aperture increased. Deep ultraviolet lithography reaches enhanced resolutions of $\geq$ 50 nm employing 193 nm wavelength light beams and, by further reducing the wavelength to 13.50 nm, extreme ultraviolet lithography achieves feature sizes as small as $\geq$ 10 nm [115, 116]. Moreover, the numerical aperture can be enhanced by using lenses with higher refractive indices or by designing complex optical systems. However, materials become highly absorptive when employing shorter wavelengths, needing the use of reflective optics rather than transmissive lenses [116].

2.3.4 Electron-Beam Lithography Patterning

Despite its inherent precision, conventional photolithography encounters a fundamental constrain in achieving patterns with dimensions smaller than approximately 100 nm due to the employed light wavelength limitation. Electron beam lithography (EBL) is used in exchange to overcome this restriction, given that the beam constituents are focused electrons with a characteristic low de Broglie wavelength, hence, increasing the patterning capabilities to resolutions $\geq$ 1 nm [117].

In theory, by following Abbe's criterion, one can calculate the resolution of an electron beam passing through a perfect optical system to be 0.12 nm at an accelerating voltage of 3×10^5 eV [118]. Nevertheless, this precision is decreased in real EBL systems by electron scattering in the exposed materials and aberrations in the electron optics, instead of by diffraction as in photolithography [119]. This challenge is ultimately determined by the point spread function of the electronic beam, typically formed by a relatively narrow peak caused by forward scattering and a broad background due to backscatter and second order effects [120]. The resolution of isolated features is limited by the width of the narrow peak. Nevertheless, the broad background component of areas away from the main beam contributes to sum up significant amounts of deposited energy. This is the so-called proximity effect phenomenon, which greatly reduces the achievable resolution. Multiple methods have been developed in order to reduce such background scattering, e.g., the introduction of a thin interlayer between the resist and the substrate that filters the backscattered electrons, a scaling of the mask patterns or the so-called ghost technique [121].

An additional improvement applicable to EBL is the patterning capability without the necessity of an optical mask since the beam can be displaced in different directions employing electromagnetic fields or electrostatic lenses [122]. Similar to optical lithography, EBL also needs positive and/or negative resist layers, but in this case, these layers are electron-sensitive. In the exposing step (Fig. 2.13b), the solubility of the resist is altered by the electron beam, and, if later developed with solvents, the exposed areas of the positive resist or the unexposed areas of the negative resist are selectively removed from the surface [122]. Upon electron exposure, the polymer chains within PMMA, a commonly employed positive photoresist [123], undergo scission. Consequently, the resulting pattern delineated on the remaining resist material mirrors the original design inscribed by the electron beam (positive pattern) (Fig. 2.13c). In contrast, a negative resist such as hydrogen silsesquioxane (HSQ) varies its solubility upon exposure, forming crosslinks that increase the resistance to dissolution of the exposed areas. This creates a negative pattern, which is the reverse image of the electron beam's original design (Fig. 2.13f) [122]. During the developing step, the resist can swell or contract, decreasing the resolution of the process [122].

Afterwards, the underlying areas of 2D material or substrate not masked by the patterned resist can be treated to achieve the desired structure. Etching is one of the most used post-lithography processes and consists of the elimination of the exposed material by aggressive techniques (Fig. 2.13d) [124]. An alternative technique involves the deposition of additional layers, typically metals or CVD-grown 2D materials, onto the regions not protected by the resist (Fig. 2.13g). The last step is the lift-off of the resist via an appropriate dissolution to leave only the final patterned structure (Fig. 2.13e and h). Notwithstanding its exceptional precision and mask-free capabilities, EBL is hampered by its slow processing speed and high cost due to its serial nature [122].

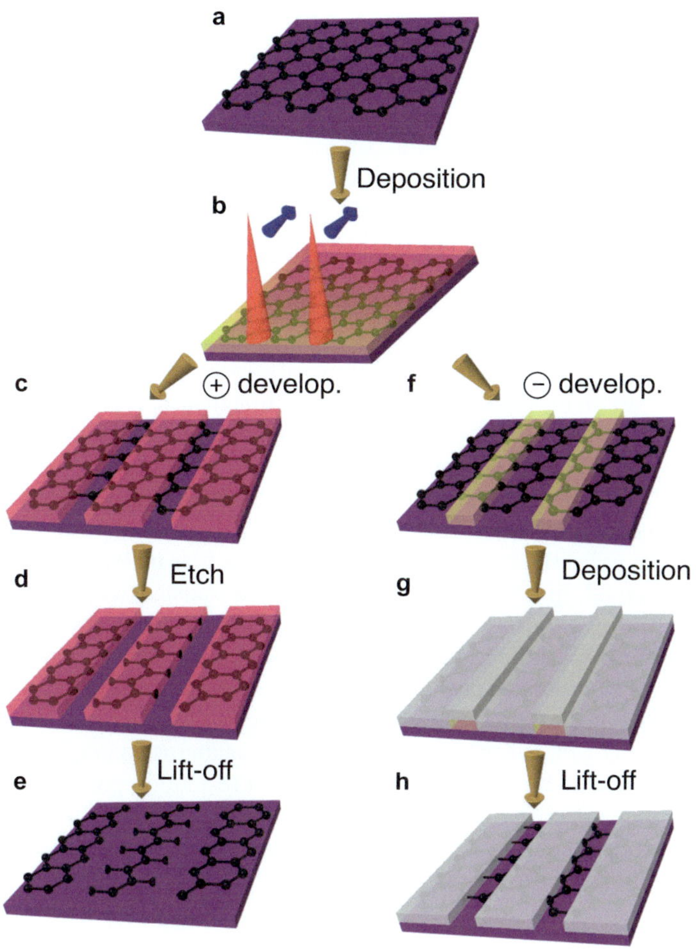

Fig. 2.13 Electron beam lithography. **a** 2D material on a substrate, **b** exposure of the ensemble to the electron beams (red cones). Positive resist developing: **c** Patterned positive resist/2D material/substrate ensemble after developing, **d** etched 2D material through the patterned voids of the positive resist and **e** patterned 2D material on a substrate after the lift-off of the positive resist. Negative resist developing: **f** Patterned negative resist/2D material/substrate ensemble after developing, **g** Metallic layer deposited on the ensemble and **h** patterned metallic layer on the 2D material/substrate structure after the lift-off of the negative resist

2.3.5 Scanning Probe Microscopy Patterning

Thermal Scanning Probe Lithography

Bombardment of the material's surface with electron or optical beam lithography has proven to be a reliable technique to obtain few-nm-sized patterns on 2D materials, but also its aggressive exposure can result in surface damage [125]. Hence, a more passive approach has been gaining importance in recent investigations.

Scanning probe lithography (SPL) utilises similar sharp probes to those employed in scanning probe microscopy, however, in contrast to mere scanning, SPL leverages the tip to inscribe high resolution patterns onto the materials while minimising structural damage.

Numerous sub-techniques have emerged within the domain of SPL, amongst them, thermal scanning probe lithography (t-SPL) stands out as a noteworthy approach. In this technique, the patterning tip is heated and rastered in contact along the target surface to inscribe a pattern via heat-induced modification of the material's structure. While t-SPL encompasses three distinct approaches, i.e., material removal, addition, and conversion [127], this review will primarily focus on the prevailing removal method. This variation induces a nanoscale excitation that cleaves the chemical bindings of the target material. The pattern's resolution depends on the tip apex diameter and the type and efficiency of tip/material thermal transport, achieving sub-10 nm patterns [127]. Therefore, temperature-sensitive materials are required to achieve the best precision [128]. Molecular glass resists and poly(phthalaldehyde) (PPA) are extensively utilised thermal responsive materials that undergo direct sublimation upon a thermal tip contact. These materials can transfer the pattern onto other substrates following exposure or directly inscribe patterns onto 2D materials, thereby yielding enhanced resolution [127]. The undoped tip is thermally activated by two resistive heaters incorporated on the cantilever. These heaters are electrically powered through heavy-doped cantilever arms connected to a power supply.

Liu et al. used t-SPL to scribe topographical nanopatterns on graphene and MoS_2 via PPA sublimation (Fig. 2.14), achieving a high resolution pattern and a strain-modulated bandgap [126]. Simultaneously to patterning, topographical information can also be extracted during the raster, monitoring in this way the real-time image of the undergoing nanoscribing [128].

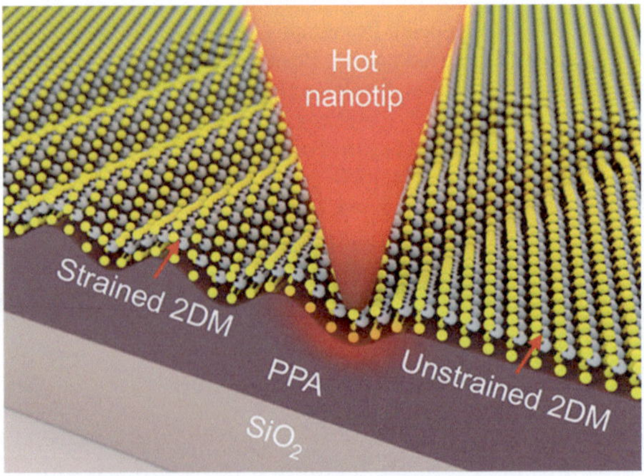

Fig. 2.14 Thermal scanning probe lithography on MoS_2. Adapted from Liu et al. [126], licensed under CC BY 4.0 [126]

Mechanical Patterning

As a straightforward but simultaneously reliable SPM lithography technique, mechanical patterning provides highly precise patterns without the need for additional intricate equipment. The procedure consists of indentation and scratching with the movable AFM tip into the target material until the desired pattern is drawn [129]. The pattern geometry is contingent upon the setpoint of indentation force exerted by the tip, which has to be balanced to be sufficient for an effective material scratching, while simultaneously avoiding excessive force that could result in the cantilever's fracture [129]. The tip apex diameter, indentation depth, tip displacement and the stiffness coefficients of both interacting structures are the critical parameters monitored to achieve a high resolution pattern.

The tip's hardness has to be higher than the target material, hence, diamond or carbon coated tips are usually employed for this purpose [130, 131]. Additionally, mechanical patterning can be combined with t-SPL to achieve a fast-rendered and highly precise pattern, resulting in a lower defect density [132].

2.3.6 Electrical Contacts to Two-Dimensional Materials

As previously elucidated in Sect. 2.3.3, a 2D device denotes a structural arrangement designed to fulfil specific physical functionalities, requiring a signal input to produce an output contingent upon the distinctive properties of the employed 2D material. In order to fabricate these devices, a bridge between the nano-environment of the 2D material and the macroscale of the entire sample is needed. For this purpose, electrical contacts, also known as electrodes, are patterned in contact with the 2D layers, linking electrically the material to external signal applications or readouts. Transistors and diodes are notable examples of these reduced-volume devices.

Electrodes are generally composed of deposited metallic materials to optimise the current flow through the connections and interfaces. In this regard, contact engineering is a key factor defining the goodness of electrical transport through metal/2D material junctions [133]. This encompasses the anisotropy of the charge carrier flow depending on the respective position between the electrode and the 2D material. In the bottom contact configuration (Fig. 2.15a), the metal is placed between the substrate and the 2D material, creating only a 2D layer/electrode vdW binding, given the lack of active covalent bonds in the cross-plane direction in these 2D flat structures. Besides the inherent Schottky barrier, an additional tunnel barrier is generated in this configuration, increasing the resistance and reducing the charge exchange at the interface [134, 135]. The same behaviour replicates in the top contact configuration (Fig. 2.15c), where the metal is deposited on top of the 2D material sitting on the substrate. However, the edge contact configuration (Fig. 2.15b) allows for an in-plane hybridisation via covalent bonds at the sides of the 2D material and the metal, avoiding the tunnelling barrier and enhancing the charge injection [133].

Fig. 2.15 Electrical contacts' configurations. **a** Bottom contact, **b** edge contact and **c** top contact configurations of metallic electrodes

Besides interface engineering, improving the charge transport in these electrodes is linked with the optimisation of the metal's work function, the doping of the contacts and the monitoring of the phase change [134]. These electrodes are generally deposited by sputtering or similar techniques, then they are patterned using lithography and ultimately transferred onto the target material via precise instrumentation [134], although the 2D material can alternatively be deposited on a substrate with prepatterned electrodes using the stamping methods depicted in Sect. 2.3.1. Other techniques employ direct growth methods such as CVD to deposit the metal on a surface and then transfer it to the material [135].

2.4 Characterisation of Two-Dimensional Nanostructures

2.4.1 General Characterisation

Morphology and Topography

Several phenomena occur on the surface of layered structures due to their interaction with the external media. The quantity of elements (atoms or molecules) that constitute the surface of nanoscale materials is much greater than the one in the bulk. These two factors contribute to generate the majority of the material's physical phenomena at the surface, becoming the paramount research location. Surface science provides insights not only to the material's external features but also to their projection on the inner layers or even the bulk.

Morphology gives a qualitative idea of the shape and the roughness of the surface elements for mutual or external comparison and identification. On the other hand, topography quantifies the components of the external layers, rendering measurements for their lateral sizes, thicknesses and intrinsic roughness. Several techniques are well suited to extract these parameters, however, AFM is likely the most employed one. All SPM techniques are characterised by using an extremely sharp tip that interacts differently with the target sample, commonly via a force or a potential that deflects a rectangular or triangular arm (called cantilever) where the tip sits (Fig. 2.16) or by direct electric, magnetic, optical or chemical tip/surface interplay without any cantilever such as in scanning tunnelling microscopy (STM).

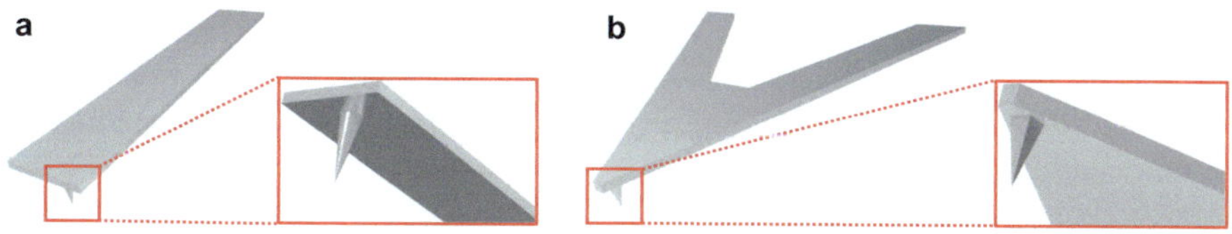

Fig. 2.16 Common SPM cantilevers. **a** Rectangular geometry and **b** triangular geometry. Insets magnify the scanning tip

Specifically in conventional AFM (Fig. 2.17) [136], the tip is in contact with the material. When the scanning is started, the repulsive or attractive forces that the material applies on the tip to oppose its movement are transmitted to the cantilever, bending it. This shifts a cantilever-reflected laser spot, allowing the laser relative movement to be read by a photodiode. The signal originating from the photodetector is controlled by a feedback loop, which corrects the position of a piezoelectric structure holding the tip or the sample, maintaining in this way a constant tip/surface force, distance or cantilever deflection depending on the scanning modality. Additionally, the feedback cycle records and converts the scanner position to render an image pixel at a particular topographical scale proportional to the tip displacement, generating topographical figures of the scanned surface area with lateral nanoscale resolution [137].

This nanoscale resolution is, in fact, one of the primary advantages of AFM characterisation. Extremely sharp probes are used to achieve lateral and vertical resolutions of 2 and 0.10 nm, respectively [138]. This allows AFM to image surface topography, measure roughness, and even detect mechanical properties such as stiffness and

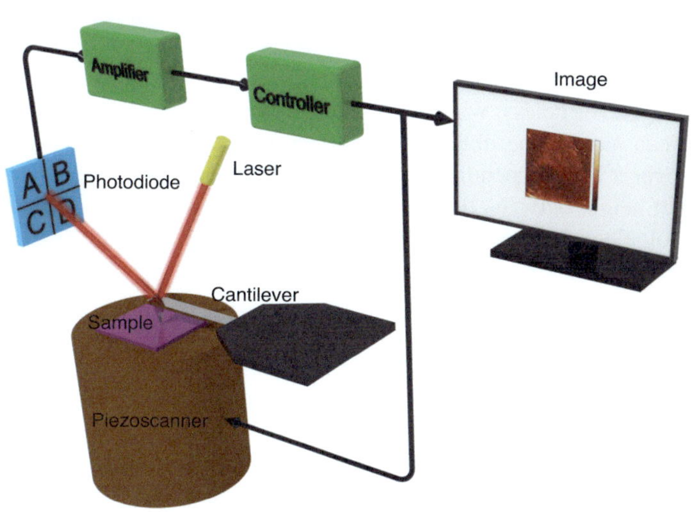

Fig. 2.17 AFM scheme

adhesion at a very small scale. Moreover, samples scanned by AFM do not require any special treatments or coatings, avoiding any potential chemical modification or damage. Additionally, AFM can operate in various environments, including air, vacuum, and liquid, making it suitable for a wide range of applications in materials science, biology, and nanotechnology.

Nevertheless, AFM also has limitations. Its resolution is constricted by the tip apex/sample surface contact radius as well as the precision of the feedback mechanism. While AFM can achieve atomic-scale resolution in ideal conditions, practical resolution is often lower due to tip wear and contamination. Furthermore, the imaging speed is relatively slow compared to other microscopy techniques. This can be a drawback when imaging large areas or in real-time scanning. Another hinderance is the potential of sample alteration due to the tip/sample interaction. In contact mode, the forces exerted by the tip can deform or damage soft samples. Even in non-contact and tapping modes, there is a certain risk of modifying the sample surface. Additionally, AFM typically provides limited information about the chemical composition of the surface, although AFM advanced methods can address this limitation.

Electronic microscopy is another widely employed technique to study surface phenomena and can be subdivided into scanning electron microscopy (SEM) and transmission electron microscopy (TEM). Both methods consist of using electron beams as the probe, directing them to impinge on the target material and be collected by multiple sensors [139]. The electrons interact and scatter differently with the varied elements in the material, transferring information about the structure arrangement and other characteristics. This can generally be read by the sensors as an energy variation of the collected electrons, being able to convert it into a morphology image. Quantitative data can also be extracted from this image to analyse the structure's thickness or roughness. SEM is preferable for studying the surface layers, while the electrons used in TEM can penetrate through the entire sample collecting data about its inner structure [140].

Optical microscopy can also provide an acceptable resolution in morphologic images [141]. Nevertheless, with the addition of profilometry, the material's thickness value can as well be obtained. This technique employs a thin diamond tip that scans over a line perpendicular to the material/substrate edge step, recording the change of height between these [142]. Moreover, spectroscopic techniques such as X-ray photoelectron spectroscopy (XPS), infrared spectroscopy (IRS) or Raman spectroscopy can quantify the structure's thickness values.

Composition

In order to obtain a complete overview of the nature of a sample, the constituent entities that comprise its structure must be analysed. This entails identifying and categorising them as either elements derived from the periodic table or molecular species. To do so, the same aforementioned techniques with further analytical methods are useful, nevertheless, most of the studies involve the use of spectroscopic techniques, given their reliability and specificity.

Raman spectroscopy is a widely applied technique in nanotechnology characterisation, providing information about the vibrational modes of the molecules that

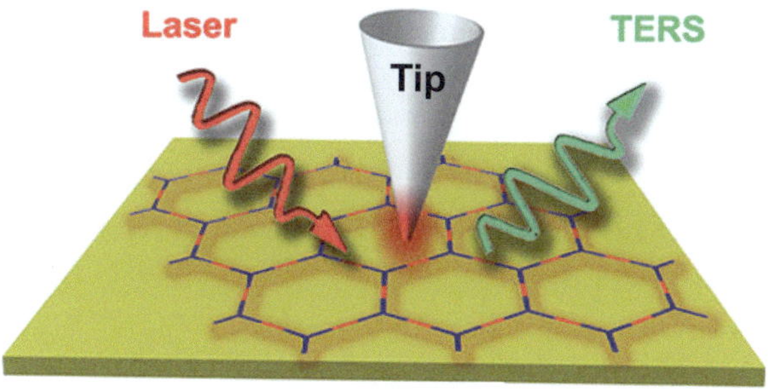

Fig. 2.18 TERS experimental geometry scheme. Adapted with permission from Ref. [146], Copyright © 2018 American Chemical Society [146]

compose the analysed specimen. A UV or visible light beam penetrates the material to interact with its vibrational energy states, causing a corresponding photonic energy exchange and frequency shift (Raman shift). Thus, the scattered photons carry the information of the interacted elements, which is collected and converted rendering a spectrum that indicates the number of photons that have interacted with a specific molecule. These data are compared to the known vibrational frequencies to identify the sample elements [143, 144]. The height of the obtained spectral peaks (intensity) represents the change in the molecular polarizability and indicates the quantity of each composing element [143]. Nearly a century has elapsed since the original Raman methodology was developed [145], this has given the research community plenty of time to engineer modifications to this technique, mainly improving the scattering signal strength and achieving nanoscale resolution [144]. Notably, it can be combined with SPM to obtain a single point Raman signal from the tip in a variant called tip-enhanced Raman spectroscopy (TERS), as seen in Fig. 2.18 [146].

X-ray photoelectron spectroscopy (XPS) is also employed for in this regard. This technique monitors the kinetic energy of the core electrons that have been ejected from the studied material (photoelectrons) after their interaction with incident photons. The energy of the photons is known, hence, by applying the energy conservation equation, the binding energy of the structure is acquired [147]. The X-ray beam is monochromatic and its photons must be powerful enough to penetrate the material [147]. In this way, information on the composition together with the electronic configuration and chemical states of the structure elements can be provided by plotting the binding energies collected by the photoelectrons. However, these are only emitted from the first layers of the material due to their ability to escape without losing all their energy, allowing to obtain valuable data of the near-surface elements, ideal for the characterisation of 2D materials [148].

As a combination of AFM and infrared spectroscopy, atomic force microscope-infrared spectroscopy (AFM-IR) is an emerging analysis technique with promising potential. In this approach, pulses of an IR laser are irradiated and tuned to the absorbance band of the target material, causing a thermal expansion, which excites the AFM cantilever up to a resonant vibration while in contact with the surface. By monitoring the amplitude of oscillation, which is directly proportional to the material's optical absorption coefficient, a spectrum can be plotted during the scan across the area of interest, rendering a fingerprint of the material's composition and obtaining enhanced-quality nanoscale measurements [149].

2.4.2 *Dynamic Nanomechanical Measurements*

Dynamic Atomic Force Microscopy

AFM can also be used to determine a broad spectrum of applied physical properties by embedding modifications to its base setup or varying its measurement process. In this sense, dynamic AFM modes have become a conventional alternative, given that nanomechanical properties can be acquired together with topography. Over the years, many dynamic AFM modes have been developed, sharing the characteristics of inducing a tip vibration via a piezo-driver contacting the cantilever and measuring the in-contact or non-contact tip/material forces at varying distances. In these modes, one of the cantilever's oscillation parameters, i.e., the frequency, the amplitude or the phase, is kept constant by the feedback loop while the surface is mapped.

Non-contact AFM utilises a vibrating cantilever to scan at a certain distance from the sample's surface, in such a manner that the tip only interacts with the material at the attractive regime of the force versus distance curve (Fig. 2.19). Although the tip apex and the material surface are close, they never engage in physical contact. When the tip's oscillation parameter is altered by a change in the attraction forces, the imaged signal also varies, capturing a map of the attraction-related mechanical features of the surface at the nanoscale [150].

Conversely, the tapping AFM mode works in a much shorter tip/sample gap, with part of the oscillation range being in the repulsive regime and tapping (in contact) on the material. The forces affecting the cantilever are regulated similarly to non-contact AFM, identifying the modulations in the amplitude or the frequency. In this near regime, it is vital to consider the effects of the humidity and the environmental contamination since the tip could form a water meniscus at the interface with the

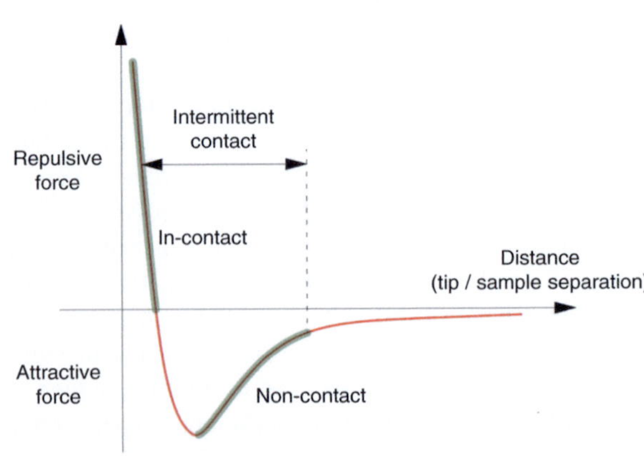

Fig. 2.19 Force versus distance AFM curve. Notice that the highlighted green area labelled as "In-contact" represents the range of the contact AFM mode, the one labelled as "Non-contact" represents the range of the non-contact AFM mode and one labelled as "Intermittent contact" represents the range of the tapping AFM mode

surface, or it could pinch pollutant particles, damaging the quality of the resulting image. The typical lateral forces present in AFM contact mode, i.e., friction and dragging, are avoided in tapping AFM thanks to the oscillation of the cantilever, which prevents continuous contact with the sample while the tip is displacing laterally, reducing the sample and tip wear [151].

Phase shift modulation is another dynamic AFM variation that correlates the phase change between the tip and driving oscillations happening when the tip encounters an area with different friction, adhesion or viscoelasticity [151].

Piezoresponse Force Microscopy

The generation of a potential difference in a particular structure due to the application of mechanical stress is the electromechanical process known as piezoelectricity. This phenomenon is reversible, being also capable of inducing intrinsic mechanical strain in a piezoelectric material by subjecting it to an electric field. An AFM variant called piezoresponse force microscopy (PFM) benefits from this effect to study the piezoelectric and ferroelectric response of nanomaterials. This technique can achieve lateral resolutions down to 3 nm and can measure buried piezoelectric features up to 1.70 μm deep [152]. Remarkably, the best sensitivity of piezoelectric surface displacement that can be recorded with PFM is 0.10 pm, allowing a notorious specificity in this regard. Interestingly, PFM can also be used to manipulate oriented ferroelectric domains.

PFM has multiple advantages, including its nanoscale resolution and picometer piezoresponse sensitivity and its ability to write and manipulate ferroelectric domains with nanoscale precision, which constitute a unique system for piezoelectric nanolithography. Like AFM, it can obtain simultaneous topographical maps to allow the deconvolution of the piezoelectric and geometrical features. However, PFM can be limited by factors such as tip/sample detrimental interactions due to tip wear and the need for specialised conductive tips. The interpretation of PFM data also requires a thorough understanding of the piezoelectric interactions between the probe and the sample, given that, in some occasions, the recorded electromechanical signal can include external contributions outside of piezoelectricity [152].

A closed-circuit electric flow is required in this system while scanning, hence, the tip has to be electrically conductive and the material must have a ferroelectric or piezoelectric nature. The converse piezoelectric effect (CPE) is the base of the process, describing how an externally applied electric field causes a deformation in the target material due to the linear interaction among the electrical and mechanical states in crystalline structures without inversion symmetry [28]. An externally modulated and lock-in amplified direct current (DC) bias is applied between the contacting tip and sample, generating an electric field that expands or contracts the measured material depending on the field direction. This mechanical deformation is sensed by the deflection shift of the cantilever. Specifically, if the initial piezoelectric polarisation of a domain is antiparallel to the electric field direction, the local area of the material will contract, decreasing the cantilever's deflection. If the electric field direction and the domain polarisation are parallel, the opposite phenomenon occurs. Thus, PFM extracts information about the direction and the relative magnitude of piezoelectric

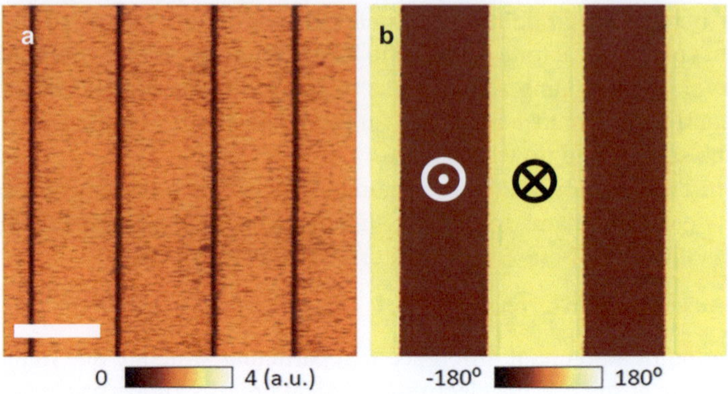

Fig. 2.20 Vertical PFM maps of a periodically poled lithium niobate (PPLN). **a** Amplitude map, showing a signal drop at the domain walls (scale bar is 5 μm) and **b** phase map, showing the contrast between positively polarised (⊗) domains and as-grown background (⊙) polarisation. Adapted from Kim et al. [155], licensed under CC BY 4.0 [155]

polarisation of the scanned domains directly from the cantilever's deflection [153]. The picoscale dimension of the piezoelectric domains causes a minimal deformation if the DC bias is not amplified, corresponding to a negligible signal-to-noise ratio (SNR). Given that such signal enhancement is performed in a lock-in amplifier (LIA) with an alternate current (AC) potential, a corresponding oscillation of the piezoelectric area at the same applied frequency is caused. The piezoelectric response is in-phase with this excitation if the domain polarisation is parallel to the electric field and out-of-phase in the opposite case, creating a 180° phase contrast between two oppositely oriented vertical domains (Fig. 2.20b). Hence, the cantilever deflection is also able to detect the vibration amplitude and phase of the domains by the demodulation of the signal received on the LIA [154, 155].

Non-vertical angled polarised ferroelectric domains can be imaged together with their specific direction information by performing vector PFM, which employs both lateral and vertical detector channels. Additionally to the deflection's track of the vertical polarisation, the lateral torsion of the cantilever is monitored to provide the in-plane polarisations of the local areas [152]. AFM and PFM can be applied to pattern by switching the ferroelectric domains' polarity with an elevated DC bias, rendering nanoscale resolution patterns at a nanosecond time lapse [156, 157].

Ultrasonic Force Microscopy

AFM can also be modulated to map elastic properties with nm precision with ultrasonic force microscopy (UFM). UFM incorporates a piezotransducer stage that induces a high vibration to the material. Thanks to this elevated frequency, much higher than the resonant frequency of the cantilever, and to the usage of a softer probe than the tip/sample contact rigidity, the cantilever becomes inertially stiff and is not able to follow the sample's vibration. This causes the tip's indentation into the material even if the latter is more rigid than the probe, measuring features hidden in the internal layers due to the non-linear mechanical detection, improving the contrast of regular force modulation or topographic measurements and probing features with different elasticity in any type of material and under multiple environments [158].

UFM is particularly useful for detecting subsurface defects or inhomogeneities up to the micron scale [159], which are otherwise challenging to observe with traditional AFM methods. Also, its lateral resolution is considered to be lower than 5 nm, comparable to PFM's specificity [160]. Further, its sensitivity is enhanced compared to other tapping techniques, given that frequency shifts are easier to accurately measure than absolute amplitudes or phase.

All these advantages come at the expense of the large required ultrasonic frequency, which can decrease the topographical resolution. Also, a previous calibration is required with materials with known elastic properties to be able to compare their UFM signal with the obtained results.

A vertical or lateral cyclic oscillation is applied from the transducer to induce a deflection or torsional excitation to the cantilever, if this resulting probe vibration is monitored, the mechanical properties of the material are obtained (Fig. 2.21) [158]. Specifically, the deflection modulates the tip indentation and the tip/surface distance, which changes alongside the sample's vibration amplitude. When the amplitude is large enough, the tip/sample contact is broken and the maximum tip/sample distance limit is set as the amplitude threshold (A_t). This amplitude is related to the slope of the corresponding force curve, which depends on the elastic properties of the material, resulting in a different type of signal for each set of mechanical properties. Thus, at a fixed adhesion force, the higher the force curve slope and Young's modulus, the smaller the amplitude threshold [161]. Although substantial forces are applied, UFM is characterised by inducing low sample damage, conversely to the classical nanoindentation, given that the tip/sample interfacial contact is interrupted multiple times due to the intense vibrational frequency, eliminating the friction force of the lateral scanning displacement [160, 161].

The data extracted from the UFM maps is directly linked with the tip/material contact stiffness (K_{Hertz}) and can be analysed to obtain mechanical features such as the local Young's modulus or Poisson's ratio of the measured specimen with the next set of equations [159]:

$$\frac{1}{a^*} = \frac{1}{a_{tip}} + \frac{1}{a_{sample}} \tag{2.2}$$

$$\frac{1}{E^*} = \frac{1-\nu_{tip}^2}{E_{tip}} + \frac{1-\nu_{sample}^2}{E_{sample}} \tag{2.3}$$

$$K_{Hertz} = 6\sqrt[3]{F_N a^* E^{*2}} \tag{2.4}$$

where a^* is a reduced radius of the tip/sample contact, a_{tip} and a_{sample} are the radii of the two spheres in contact, i.e., the tip apex and the material, E^* is the reduced Young's modulus, E_{tip}, E_{sample}, v_{tip} and v_{sample} are the tip and sample Young's moduli and Poisson's ratios, respectively, and F_N is the normal contact force.

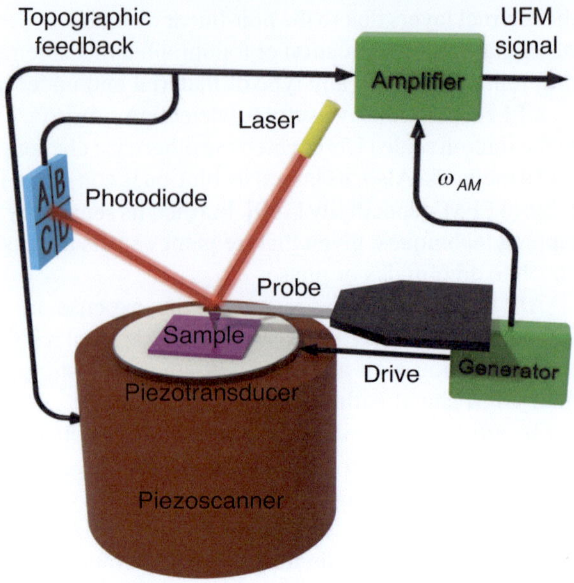

Fig. 2.21 UFM scheme. Notice that ω_{AM} denotes the amplitude-modulated frequency input into the amplifier

Graphene and MoS_2 have been widely studied employing UFM. Robinson et al. determined the nanomechanical properties of suspended MoS_2, eliminating the sample/substrate interaction. They found a clear transition between the suspended and supported monolayer and the thickness dependence, originating from a higher signal variation for thinner flakes, which were softer according to the darker extracted UFM contrast [163]. Skilbeck et al. probed the mechanical properties of graphene islands on a flat copper surface [162], identifying them thanks to the higher contrast provided by the UFM in comparison to the topography (Fig. 2.22). In addition, the softness of graphene compared to the copper substrate is deduced from the same image. This result can be attributed to the soft nature of the interface between the two materials, decoupling the graphene from the substrate to be exclusively physisorbed onto it.

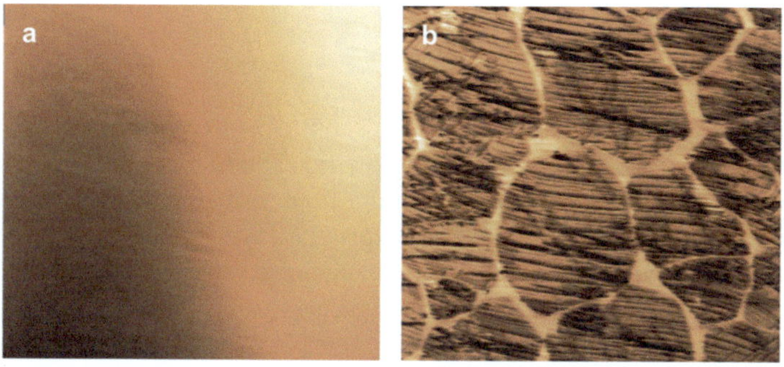

Fig. 2.22 $10 \times 10\,\mu m^2$ **a** topography and **b** UFM images of graphene islands on a copper substrate. Adapted from Skilbeck et al. [162], licensed under CC BY 3.0 [162]

Heterodyne Force Microscopy

Subsurface imaging of nanomechanical properties can also be achieved with heterodyne force microscopy (HFM) given its characteristic ultrasonic modulation. Even though its original implementation only achieved qualitative observations, further developments unravelled and quantified the dynamic behaviour of the analysed samples with a lateral resolution lower than 10 nm [164].

The HFM setup consists in a double piezotransducer ultrasonic MHz excitation at the probe and the sample, however, the frequencies differ slightly in the kHz range. Analogous to UFM, such frequencies are much larger than the first resonance frequency of the cantilever. Hence, when the tip and sample are in contact and the scan is initiated, multiple nonlinear tip/sample interactions occur, consisting of attractive vdW and repulsive Hertzian contact forces. Such interaction generates a heterodyne low frequency signal at the difference mix of the ultrasonic excitations, which is generally modulated to be smaller than the resonance frequency of the cantilever, ensuring that the cantilever oscillates at the difference frequency. By monitoring the amplitude of such difference beating frequency, one can obtain nanomechanical properties such as adhesion and deformation, while the phase map provides information about the time resolve properties, e.g., the dynamic relaxation or the viscoelasticity [165].

Considering the low frequency terms acting on the cantilever's response, the force that it receives from the sample (F_N) can be expressed as [165]:

$$F_N = \alpha_{nl}\left[\frac{A_1^2 + A_2^2}{2} - A_1 A_2 \cos(\omega_b t - \omega_2 \zeta)\right] \tag{2.5}$$

where α_{nl} is a nonlinear coefficient, A_1 and A_2 are the tip and sample amplitudes, respectively, ω_b and ω_2 are the beating and sample frequencies, respectively, t is the time and ζ is the viscoelastic relaxation time. The $\omega_2\zeta$ phase term represents the dynamic mechanical phenomena on the sample surface, preserving its dependence at high frequency. Thus, monitoring the amplitude and the phase of the signal by a LIA can be performed at the beat frequency.

2.4.3 Electrical Characterisation

Kelvin Probe Force Microscopy

Kelvin probe force microscopy (KPFM) allows to capture maps of the contact potential difference (CPD) and the work function of the probed element's surface [166]. KPFM's capabilities are limited to the superficial layers of the structure since the CPD and the work function are surface-related characteristics [166].

Analogous to dynamic AFM, KPFM operates in the tapping mode, with the cantilever oscillating at its resonant frequency [167]. The probe must be electrically conductive to sense the tip/sample CPD, requiring a conductive coating. This coating (which also increases the resistance to wear of the probes) can be composed of doped diamond, platinum silicide (PtSi) or platinum/iridium, similar to the ones used in scanning spreading resistance microscopy (SSRM) [168, 169]. Once the measurements are started, an AC bias is applied to the probe, generating an electrostatic force that can induce an additional vibration on the cantilever. To compensate for this effect, the force is controlled by a LIA, which supplies a DC voltage to nullify the additional oscillation. When the tip encounters a sample's area at a different surface potential, the recorded electrostatic force alters, causing a corresponding shift to the mentioned cantilever's oscillation that has to be corrected by also varying the monitored DC bias, obtaining in this way a quantitative image of the surface's CPD [166, 170, 171]. The CPD originates from the work functions of each interacting element, i.e., the tip and the sample's surface, hence, it is possible to qualitatively map the work function difference between these two entities. However, to quantify the local work function values of the material of interest, a model that describes the electrostatic tip/material interactions and the knowledge of the tip's work function value is required [170, 171].

KPFM has two modes of operation, i.e., the single and dual pass setups. Their distinction lies in the nomenclature, as the dual pass entails two successive scans over the same fast-direction virtual line, while the single pass involves only one. The single pass configuration comprises a set of advantages by concurrently recording the topography and the CPD, for instance, improving the correlation between the two signals, having a quicker scan ratio and reducing the tip's wear. Also, higher sensitivity and KPFM signal resolution are achieved in this setup due to the close tip/sample distance originating from the small amplitude in the cantilever's vibration, although the spatial resolution can be reduced and the edge artefact effects are increased [172]. Dual pass KPFM renders the topography on the first scan with a tip/material oscillatory contact regime (Fig. 2.23a). In the subsequent scan, the tip is displaced away from the sample's surface, usually tens of nm, and the AC and DC voltages are applied, recording the CPD (Fig. 2.23b). This distance has to be engineered to avoid a tip/material collision or a stray capacitance from the cantilever to the sample. This setup provides a superior spatial resolution and lower noise in the images. Since topography is already known from the first scan, the tip is able to maintain a constant separation with the material on the second pass, avoiding edge and boundary effects [172].

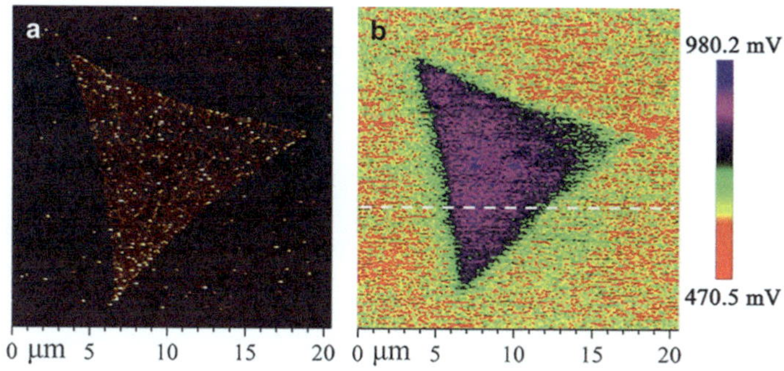

Fig. 2.23 **a** Topography and **b** KPFM CPD maps of perovskite quantum dots (PQDs) on a MoS_2 monolayer. Adapted from Wu et al. [167], licensed under CC BY 4.0 [167]

Equations (2.6)–(2.9) quantify the cantilever's response to the input potential. The parabolic relation between the electrostatic force (F_E) and the tip's bias (V_{tip}) is [172]:

$$F_E = -\frac{\nabla C_Z}{2}\left(V_{tip} - V_{CPD}\right)^2 \tag{2.6}$$

where ∇C_Z is the tip/sample capacitance gradient with respect to the separation distance (Z) and V_{CPD} is the CPD. The second set of equations demodulates the forces acting on the cantilever and relates them to the electric potential:

$$F_S = -\frac{\nabla C_Z}{2}\left[(V_{DC} - V_{CPD})^2 + \frac{V_{AC}^2}{2}\right] \tag{2.7}$$

$$F_\omega = \nabla C_Z (V_{DC} - V_{CPD}) V_{AC} \sin(\omega t) \tag{2.8}$$

$$F_{2\omega} = \frac{\nabla C_Z V_{AC}^2}{4} \cos(2\omega t) \tag{2.9}$$

where F_S is the static force, F_ω and $F_{2\omega}$ are the first and second harmonic components of the dynamic force, respectively, V_{DC} and V_{AC} are the applied DC and AC voltages, respectively, ω is the frequency and t is the time.

Scanning Gate Microscopy

Scanning gate microscopy (SGM) incorporates a conductive tip utilised as a movable gate that exhibits a capacitive coupling with the sample and, given the local influence of the probe over its electrostatic transport [173], maps the material's electrical conductance as a function of the tip potential and position.

Generally, the top gate tip is biased with an AC voltage and brought to close proximity to the target material, altering its charge carrier concentration and electrostatic potential during the scan over a conductive area. At the same time, a DC biased

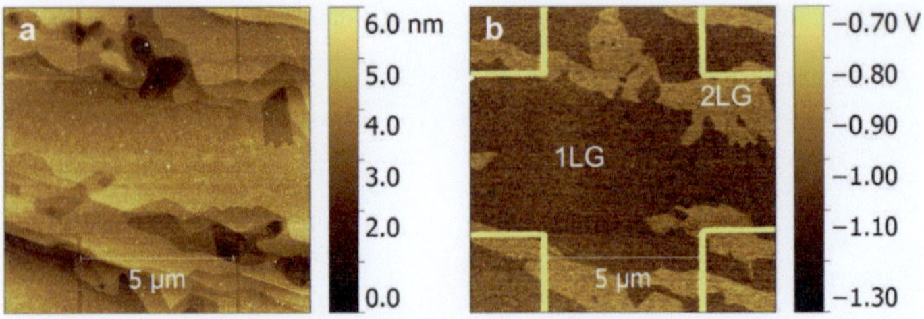

Fig. 2.24 **a** Topography and **b** SGM surface potential maps of mono- and bilayer graphene. Adapted from Mackenzie et al. [174], licensed under CC BY 3.0 [174]

current is applied to the sample via metallic electrodes [174]. If the AC voltage shifts across the sample (originating from the varied charge density) as a function of the tip position is measured, a map of the conductance versus the applied tip voltage can be acquired together with the surface topography (Fig. 2.24) [173]. This provides a localised insight into how the charge carriers organise and how they flow within the material after a given potential perturbation. SGM can also probe buried features in the near-surface layers given its tip/sample capacitive coupling nature [175].

If a cryogenic environment under the mK regime is subjected to the sample while performing SGM, quantum scale maps of the electron mobilities can be rendered, given that their mean free path increases inversely to the temperature. Moreover, quantum-based electronic conductance anomalies can be monitored in situ and at a nanoscale resolution [176].

2.4.4 Thermal and Heat Transport Characterisation

Scanning Thermal Microscopy

Most thermal characterisation techniques lack area specificity or localisation. Scanning thermal microscopy (SThM) is an AFM-based technique that is able to quantitatively obtain thermal features of materials alongside topography with high sensitivity and spatial resolution. Specifically, SThM's purpose is to probe the temperature shift in a material's section or its thermal conductivity. Furthermore, with analytic models and calibration, SThM provides insight into other heat transport features such as thermal diffusivity, heat capacity, glass transition temperature, latent heat, enthalpy, etc.

Besides the simultaneous thermal and topographical rendering capability which causes SThM to share the same benefits and drawbacks as contact AFM, its main advantage is the achievable sensitivity (ultimately limited by the circuit noise) and lateral resolution, which can reach up to the order of the 15 mK and 10 nm, respectively [177]. However, the technique can be challenging to calibrate, and the thermal probe might disturb the thermal field of the sample under ambient conditions, which can limit the accuracy of measurements.

Although in the SThM configuration additional electronic equipment is added compared to the usual AFM system, the paramount element is still the probe. In order to obtain topographic figures, the cantilever and the tip remain as the regular scanning deflection system in contact mode, however, a temperature-sensitive heater, also known as resistance thermometer, or a thermocouple is incorporated into the probe's structure to extract the thermal properties. At the same time, the probe is monitored from a Wheatstone bridge (BRG) as its feedback source by balancing the bridge voltage to either measure or maintain a constant probe temperature (Fig. 2.25) [178].

SThM also has two different variants. The passive mode is applied to extract the temperature changes in the sample. In passive SThM, a prior calibration of the tip parameters has to be performed, doing so by subjecting it to thermal equilibrium with a reference material with known heat characteristics, establishing its heat resistance and eliminating the potential difference in the BRG. Following the calibration, the tip is brought to contact with the target material to start scanning, exchanging heat by conduction or convection depending on whether there is a pure tip/sample physical contact, a water meniscus or air contact in humid or ambient environments, respectively. Ultimately, this thermal transfer depends only on the material's heat behaviour, given that the tip influence is known due to the calibration and the interface effects are disregarded via subsequent analytical treatment or by performing the measurements under controlled environments. Due to the heat exchange, the temperature of the tip varies, causing a change in its resistance related to the shift of the bridge's voltage balance. By knowing the constant probe current and monitoring the voltage shift, Ohm's law can be applied to obtain a map of the local temperature change. To avoid probe self-heating, the current has to be small, otherwise, an additional resistance shift could be generated externally from the target thermal results [179, 180].

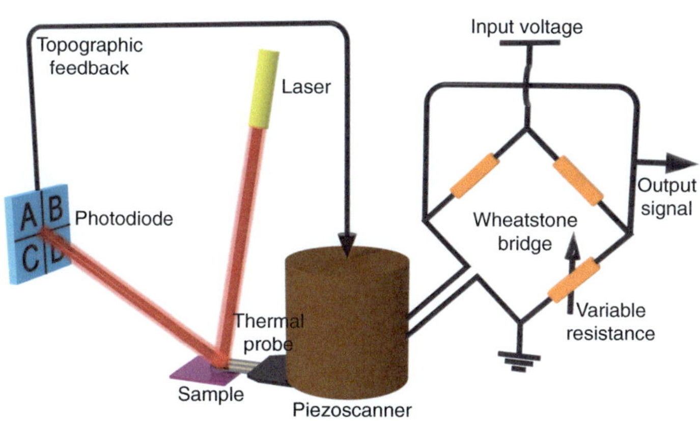

Fig. 2.25 SThM scheme

Active SThM is primarily employed to measure thermal conductance values by utilising the tip's resistance as a controlled constant temperature heater via the feedback loop. Scanning begins after setting the tip at a constant temperature higher than the material, imposing a heat transfer from the two media which cools the tip. The temperature loss is controlled by the feedback loop shifting the voltage of the BRG until the specified tip's temperature is recovered and the original resistance is restored. By mapping this voltage variation, the thermal conductance features are quantified. The signal can have thermal data from three different structures, i.e., the heat flow in the tip, the tip/sample interface (related to the tip/sample contact area), and the sample itself (including the top layer, the sample/substrate interface and the substrate heat transport) [179, 180].

The initial signal acquired from the SThM carries the thermal transport information of the scanned structure, however, a quantification model capable of transforming these data into thermal conductivity values or other heat parameters is required. Fourier's law is useful in this regard since it depicts an expression that relates the thermal transport with any heat transfer direction inside the material [181]:

$$\Phi_{th} = \kappa \frac{\Delta T}{l} \tag{2.10}$$

where κ and l are the thermal conductivity and the length in the heat flow direction of the material, ΔT is the temperature difference and Φ_{th} is the heat flux. Quantifying the heat resistance created in a geometrical constriction inside a certain material is a matter of interest. Thus, the next equations are used to relate the diffusive and ballistic constriction thermal resistances (R_{Cd} and R_{Cb}, respectively) with the constriction's dimension [182]:

$$R_{Cd} = \frac{1}{2\kappa l_C} \tag{2.11}$$

$$R_{Cb} = \frac{4\lambda_L}{3\kappa l_C} \tag{2.12}$$

where λ_L and l_C are the material's respective phonon mean free path and constriction length.

Due to the nanoscale capabilities of SThM, where quantum behaviour starts to influence the thermal transport, it is useful to understand how heat behaves at such small dimensions. Heat transport in the solid-state is a complex process where thermal energy is transferred through atomic interactions and movement of charge carriers. Heat energy is therefore carried by both phonons and electrons. The former are quantised lattice vibrations constituting the primary heat carriers in non-metallic solids and insulators. Their behaviour can be described by the phonon Boltzmann transport equation, which considers scattering processes and phonon/phonon interactions [183]. Phononic thermal conductivity is influenced by the material's atomic structure, bonding, and defects, which affect the phonon mean free path and lifetime.

In metallic solids, electrons are the dominant heat carriers. The electron contribution to thermal conductivity can be described by the Wiedemann–Franz law, which relates the thermal and electrical conductivities. Electrons, being fermions, follow Fermi–Dirac statistics, and their heat transport is influenced by the electron DOS and scattering mechanisms, such as electron/phonon interactions [184].

At the quantum scale, the physics of heat transport becomes more nuanced. Phonons, as bosonic quasiparticles, interact via scattering events, which can be influenced by temperature, impurities, and grain boundaries. These interactions determine the phonon mean free path and, consequently, the thermal conductivity. For electrons, quantum effects such as quantum confinement and tunnelling can significantly alter their transport properties, especially in nanostructures and low-dimensional systems [185].

Although in solid-state materials conduction is the main influencer in heat transport, thermal behaviour can also be affected by convection and radiation to a certain degree. Convection involves heat transport in weakly interacting fluids, e.g., gases or liquids, however, especially in low-dimensional materials where surfaces and interfaces become crucial, this type of thermal transport must be considered when measuring the heat properties of a solid that is not isolated from surrounding media like air or other gases, e.g., in air-SThM. In the other hand, thermal radiation, described by Planck's Law, can become important at high temperatures but is often a secondary effect compared to conduction [183, 186].

Finally, it is worth commenting on a crossover between the 3ω-method and SThM known as 3ω-SThM. In this procedure, an AC biased current is applied on the tip's resistor with a frequency ω, causing Joule heating at a 2ω frequency, i.e., the second harmonic, and promoting the tip's resistance to the same frequency. The measured voltage shift at the tip's resistor corresponds to the product of the 2ω resistance and the ω AC biased current, resulting in a 3ω component directly related to the probe's temperature. This 3ω signal is measured by the LIA with the SThM BRG, rendering 2D maps of the temperature and the thermal conductivity with improved spatial resolution and thermal specificity [178, 187].

Scanning Thermal Gate Microscopy

Scanning thermal gate microscopy (STGM) has recently arisen as a process that combines the approaches of SThM and SGM. STGM maps the 2D structures' thermovoltage and their local Seebeck coefficient with a lateral resolution of a few tens of nm and μV sensitivity [188]. Contrary to some SPM techniques, STGM can measure the local thermoelectric properties of 2D materials without damaging their structure or modifying their properties. For instance, adapted SThM can measure the local TE properties, but it requires an electrical contact between the tip and the sample, changing the TE behaviour of the latter, while STGM does not require this, eliminating the linked uncertainty. STGM also excludes the phase-driven effects, increasing the quality of the local Seebeck coefficient measurements through signal deconvolution and avoiding the long integration times characteristic of the use of various LIAs in SThM TE analysis, prompting quicker scanning speed and making STGM appliable to an SThM setup [188].

The tip in the STGM configuration is heated by an elevated AC voltage for the thermal conductance measurements and biased with a DC offset voltage for the temperature sensing, causing Joule heating to increase the tip's temperature to values hotter than the sample and starting the heat transfer towards the material. When the hot tip scans over the material, a local symmetric temperature gradient is caused around the contact area and the overall TE voltage drop in the device is measured with electrical contacts on the sample edges, obtaining the thermovoltage values following the Seebeck effect [188].

Due to the novelty of this technique, few studies have obtained thermoelectric parameters employing it. However, this results in a potential pathway for the modification of STGM to overcome drawbacks like the required precise temperature control of the probe as well as the high input voltages for its heating.

2.5 Physical Phenomena in Two-Dimensional Materials and Devices

2.5.1 Heat Transport and Advanced Thermal Phenomena

Heat Transport Anisotropy

By definition, a material becomes anisotropic when its intrinsic quantifiable properties are conditioned by the measured direction, making them orientation dependent. This is a classic behaviour of crystalline structures due to their diverse arrangement of atoms. Hence, vdW materials present a notable anisotropy in their physical characteristics not only due to their different crystalline configurations but also in light of their minimal thickness, vdW interplane bonding and interface proximity. In fact, the properties characterised at the cross-plane direction generally present a higher degree of anisotropy when comparing them to their in-plane analogue than when the two in-plane anisotropic directions (A-C and Z-Z) are contrasted. This is because the in-plane A-C and Z-Z anisotropy occurs only due to crystalline arrangement change, while in the cross-plane equivalent, the high surface/bulk ratio in 2D materials increases the effects of the interfaces to other media.

Specifically, thermal transport properties are greatly affected by the anisotropy of layered structures, providing researchers with another tuning tool for the implementation of such materials in real-world devices. For instance, a certain 2D material could be used in heat management applications benefiting from its generally high in-plane thermal conductivity or exploited in thermoelectric devices hindering the heat transport by forcing it through the cross-plane direction [189, 190]. Nevertheless, due to the complexity of measuring the thermal properties of atomically thin materials in the cross-plane direction, most of the studies focus just on their in-plane anisotropy [191].

Table 2.8 Thermal conductivity anisotropic ratios of different 2D materials

	Graphene	MoS_2	InSe	h-BN	References
$\varphi_{\parallel}$	1.01	0.80	1.01	0.47	[84, 192–194]
$\varphi_{\perp}$	666.67	17.33	11.18	400	[85, 90, 92, 94]

Additionally, the property that depicts heat transport anisotropy in the most effective and understandable way in 2D materials is their thermal conductivity. In Sect. 2.2.4 it was possible to elucidate the quantitative values of the different thermal conductivities of graphene, TMDs, InSe and h-BN in their different directions, which provides the opportunity of extracting each one of their approximate anisotropic ratios as:

$$\varphi_{\parallel} \equiv \frac{\kappa_{A-C}}{\kappa_{Z-Z}} = \frac{\kappa_x}{\kappa_y} \tag{2.13}$$

$$\varphi_{\perp} \equiv \frac{\kappa_{in-plane}}{\kappa_{cross-plane}} = \frac{\kappa_{xy}}{\kappa_z} \tag{2.14}$$

where $\varphi_{\parallel}$ and $\varphi_{\perp}$ are the A-C (κ_x) versus Z-Z (κ_y) parallel directions and the in-plane (κ_{xy}) versus cross-plane (κ_z) perpendicular directions thermal conductivity ratios, respectively. Hence, summarising them in Table 2.8.

As demonstrated in Table 2.8, the cross-plane versus in-plane anisotropic ratio values are much higher than the in-plane A-C versus Z-Z analogous, which are close to unity. This simultaneously proves two diverse phenomena. First, a clear similarity between the thermal conductivity values in the A-C and Z-Z directions is observed in 2D materials due to the minor change in crystallographic orientation for each direction and the shared covalent arrangement between the composing elements, increasing the quality of the heat transport. Secondly, the vdW interlayer bonding, the decrease of material concentration and the interphase proximity diminish the cross-plane thermal transport behaviour in all the layered structures, ultimately rising the perpendicular thermal conductivity ratio [190]. Interestingly, the phenomena that cause thermal anisotropy are still a matter of debate in the scientific community, however, multiple proposals have been published in the literature. Amongst them, the thickness, the lateral dimensions and the composition are the utmost contributors, however, other factors such as the lattice alignment, the interlayer twist angle or the defect concentration can also play a role in the heat flow trajectory [189, 190, 195].

Substrate Interaction

Two-dimensional materials present a lack of stability if the conformity with the surface on which they are deposited is not adequate, given their thin and flexible nature, resulting in the formation of ripples or curls. Hence, the epitaxy to the holding substrate is vital for maintaining the structure's intended performance. However, by introducing an additional material to the ensemble, a corresponding interface is created, affecting the system's properties. In thermal transport, this interface acts as an additional heat resistance, quantifiable by the interfacial thermal insulance (r_{int}). The majority of substrates interacting with a 2D material are solid structures, however, layered entities can also have a proper epitaxy with gel-like and liquid systems [196, 197]. There is even the possibility of eliminating the substrate's necessity at certain configurations by magnetically levitating the 2D layer [198]. The 2D material/solid substrate interaction can be divided into two configurations, in the first one, the entire layer rests flat on the substrate, and in the second one, a region of the 2D material is not in direct contact with the supporting surface, having both of its faces exposed to the external media, these configurations are called supported and suspended, respectively.

The supported 2D material configuration is the most common substrate interaction in the field, given that no previous patterning treatment is needed on the substrate and the transfer is straightforward. Despite these, the thermal properties will be largely influenced by the interface given the bigger contact area, causing an increase in the cross-plane thermal conduction from the sample to the substrate, as depicted in Fig. 2.26a, while the total conductivity is compensated with a reduction of the in-plane heat transport [199]. In contrast, suspended samples only interact with the substrate at certain areas, as shown in Fig. 2.26b, enough to avoid the vdW material plunging into the hole-like free space. In this suspended region, the cross-plane thermal transport is suppressed, given that most of the layered area has no solid/solid interface to allow conduction. A characteristic drawback of this layout is the requirement of substrate pre-lithography, which implies a time-consuming process. Nonetheless, the characterisation performed on the resulting suspended samples possesses the virtue of isolating the properties of only the 2D material, given that the influence of the substrate and the interface is eliminated. Thus, the thermal properties of the suspended materials acquire a higher significance in the in-plane direction, enabling their characterisation together with the effects of convection and radiation to the medium.

Wang et al. obtained the in-plane thermal conductivity of FLG suspended and supported on SiO_2 using a thermal bridge configuration [200]. They found that the substrate reduced the thermal conductivity of graphene in the supported arrangement, nevertheless, as the measured layers were further from the substrate, this

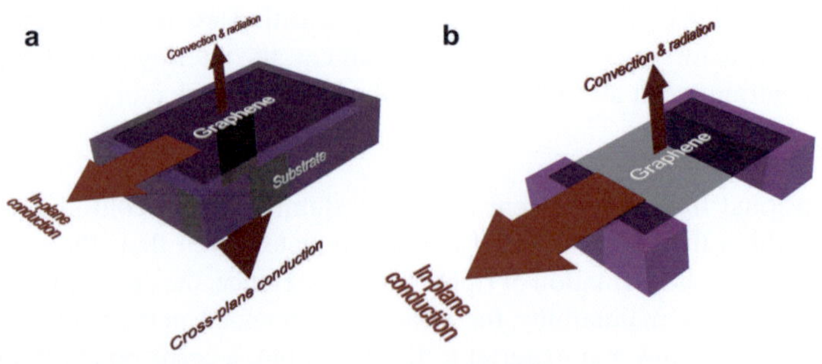

Fig. 2.26 Thermal transport in **a** supported and **b** suspended graphene

lowered heat conductivity effect diminished. At the suspended configuration, they plotted a temperature versus thermal conductivity dependence following a power law, attributed to the significant contribution of the flexural phonon modes to the heat transport. Interestingly, the extracted thermal conductivity values of the suspended graphene were smaller than the ones predicted in the theory due to the phonon scattering at the boundaries of the graphene/substrate contact regions. A similar study of supported graphene's thermal properties was performed by Kim et al. employing optothermal Raman spectrometry [199]. They presented a dataset expressing the variation of graphene's thermal conductivity with respect to the applied temperature. Additionally, by skipping the annealing step, micro-suspended areas were created due to the intrinsic substrate's roughness, allowing to compare the thermal transport with the fully supported structure.

The influence of substrates in TMDs has also been a matter of study. For instance, Zhang et al. used an equivalent optothermal Raman method to obtain the values of the interfacial thermal insulances in supported and suspended MoS_2 and $MoSe_2$ configurations as well as their lateral thermal conductivities [100]. Whereas Carvalho et al. applied the same technique to characterise the temperature-dependence of the cross-plane A_{1g} Raman mode of suspended and supported monolayer WSe_2 [201]. Both studies showed a clear dependence on the substrate configuration linked with its roughness and interface.

Heat Transport in Two-Dimensional Heterostructures

The assembly of vdW heterostructures is a distinctive approach to modulate the properties of 2D systems by stacking two or more 2D layers. Hence, the thermal transport also varies for the array of piled layers compared to their isolated counterparts. For instance, the scattering of heat carriers in heterostructures implies additional dissipation sources between the layers, causing Joule heating. Offering a high tunability, these stacked ensembles can be utilised in heat applications that require elevated or diminished thermal conductivity values, i.e., electronic devices or thermoelectric systems, respectively. The capability of heterostructures to introduce defects, interfaces, stress or structural diversity generates a wide range of possibilities to control their thermal phenomena and acquire a deep insight into their underlying mechanisms [202, 203].

The in-plane thermal conductivity of graphene supported by h-BN was calculated using MD simulations by Zhang et al. revealing a reduction of the overall thermal conductivity of the system by 23% when compared with the 1LG suspended structure, however, the thermal conductivity of the graphene/h-BN supported structure was still greater than the 1LG/SiO_2 supported system, suggesting an improvement in heat dissipation when the h-BN was employed. Moreover, they outlined that the heat transfer characteristics in the heterostructure were strongly dependent on the stacking fashion of the layers [204]. Wang et al. executed the atomistic Green's function method to obtain the thermal conductance of graphene coupled with h-BN [205]. They stated a decrease of this property at low temperatures as well as its possible preservation as the 96% of the thermal conductance in RT graphene, given that the h-BN coupling affected the cross-plane vibrations generating linear flexural phonon modes and eliminating their transmittance. Raman spectroscopy measurements were conducted by Chen et al. on a graphene/h-BN heterojunction to obtain an interfacial thermal conductivity value of 7.40 MW m^{-2} K^{-1} at RT [206].

Other 2D materials have also been combined to extract their homo- or heterostructure's thermal properties. Boltzmann transport equation was applied by Ma et al. to derive the RT thermal conductivity of a $MoS_2/MoSe_2$ bilayer heterostructure, resulting in a value of 25.40 W m^{-1} K^{-1} [207]. Furthermore, they showed an inverse dependence in the sample temperature versus the in-plane thermal conductivity, also linked with an increase of the interfacial thermal conductance at lowered in-plane heat transport. Gao et al. observed a decrease in the in-plane thermal conductivity of graphene supported on MoS_2 due to the interlayer coupling, remarkably, this reduction did not affect the heat transfer in the TMD [208]. Finally, Liu et al. made use of optothermal Raman to plot the interfacial thermal conductivity of a MoS_2/h-BN heterojunction, obtaining a saturation at a value of approximately one-third of the graphene/h-BN's ensemble for temperatures higher than 100 K.

2.5.2 Multiphysics Phenomena in Two-Dimensional Materials

From Dimensionless to Bulk

Two-dimensional materials can be modified by top-down or bottom-up processes to remove or add spatial dimensions, respectively. In other words, structures with zero, one and three dimensions are producible from a 2D sample. As described in the previous sections, transitioning from a bulk crystal to its 2D form entails new physical phenomena, thus, by furtherly decreasing to singular dimension and dimensionless sizes, novel properties are also unveiled, given that, in each dimension reduction, the quantum influence becomes more important.

In general terms, a nanoparticle is a structure with all of its three spatial dimensions reduced to a nanoscale size, considering it non-dimensional, i.e., 0D. Nanoparticles can be used as building blocks to assemble higher-dimensional structures via bottom-up approaches [209, 210] or introduced in vdW systems to enhance their properties [211, 212]. Reversely, 2D-based nanoparticles can be formed starting from layered materials with top-down approaches [213, 214]. Quantum dots (QDs) are the clearest example of nanoparticles (Fig. 2.27a). If derived directly from a 2D material, they can show a plethora of innovative features. For instance, graphene QDs are fluorescent and semiconductor materials with size-dependent bandgaps, different to their zero-bandgap semimetallic behaviour in their 2D counterpart. These characteristics arise from the imposed quantum confinement of zero-dimensionality [215]. On the other hand, TMD QDs are characterised by their high effective electronic mass in the valence and conduction bands, exhibiting substantial exciton binding energy and pronounced excitonic effects at RT attributed to the amplified Coulomb interactions in the spatial confinement and limited dielectric screening, generating multiple excitonic forms [216]. h-BN is an electronic insulator in its two-dimensional form, however, when decreased to the QD dimension, its bandgap can be reduced thanks to edge hydrogenation or doping, improving its charge transport. This bandgap can be furtherly tuned to achieve high absorbance in the UV regime and emit blue light as a weak fluorescent QD [215].

Although many nanoscale systems can be engineered in the singular dimension configuration, e.g., nanowires, nanorods, nanofibers and nanobelts, the ones that can be derived from a 2D system are nanotubes and nanoribbons (Fig. 2.27b and c, respectively). In both cases, the thickness and width are restricted to the nanoscale, i.e., < 100 nm, imposing quantum confinement, while the length is the only allowed macroscale dimension (1D). 2D-based nanoribbons are mostly prepared by employing bottom-up techniques, and their properties can be significantly modified by inducing strain, doping or phase engineering [217]. Given the difference between the width and the length, the behaviour of 2D nanoribbons varies depending on the width-confined arrangement (A-C or Z-Z). For instance, A-C graphene nanoribbons show metallic or semiconductor properties, whereas their Z-Z analogous are limited to metallic behaviour given their much lower bandgaps. This type of nanoribbon is also a widely flexible structure and a proficient heat conductor. MoS_2 nanoribbons are also direct bandgap semiconductors [217]. However, their more energy-favourable configuration is the nanotube curled arrangement due to its low surface energy and strong attraction covalent bonds [218]. This configuration can be used as a template for growing smaller nanostructures such as nanowires or

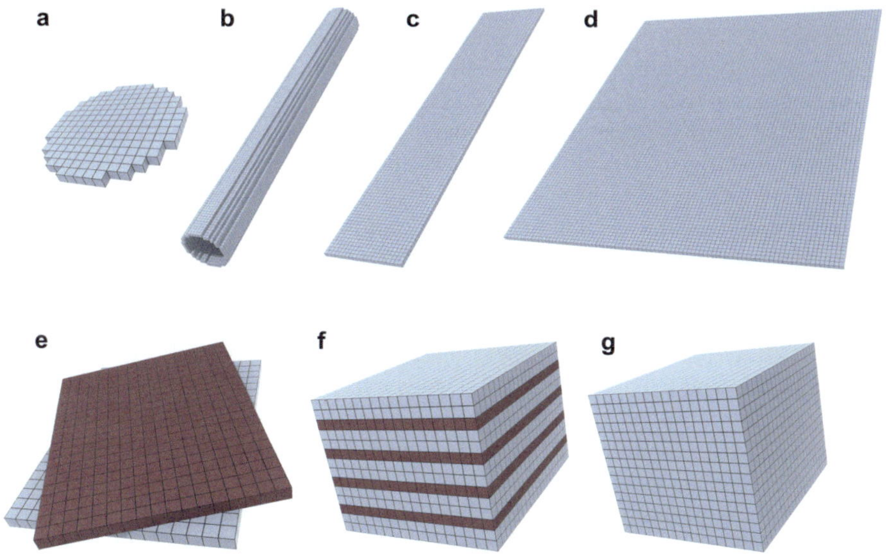

Fig. 2.27 Different dimensions of 2D-based structures. **a** Quantum dot (0D), **b** nanotube (1D), **c** nanoribbon (1D), **d** nanosheet (2D), **e** twisted heterostructure (2D), **f** vertical superlattice (3D) and **g** bulk (3D)

coaxial nanotubes [218]. Graphene-based carbon nanotubes (CNTs) are flexible and outstanding thermal and electrical conductors, while TMD-based nanotubes outperform the electronic properties of their layered counterpart and boron nitride nanotubes (BNNTs) provide directional electronic insulation while showing an efficient heat conductivity [219–222].

2D systems emerge in various shapes, being nanosheets, also known as nanolayers, the most common simplified configuration with only its thickness restricted to the nanoscale (Fig. 2.27d). These nanosheets can be stacked over each other to form homo- or heterostructures, with their corresponding novel physical qualities. Although being embedded in a three-dimensional space and exhibiting some 3D features, these stacked structures are generally considered 2D systems since their thicknesses are still much smaller compared to their lateral dimensions, maintaining quantum confinement in the perpendicular direction.

An additional tool for heterostructure modification is studied in the field of twistronics, where the layers of the stack are rotated at different angles with respect to each other (Fig. 2.27e). The importance of this field has exponentially risen since the discovery of superconductivity and correlated insulating states in a twisted graphene heterostructure at a "magic" angle in 2018 [223, 224] Artificial solid materials of this nature exhibit atomically precise interfaces devoid of interdiffusion. Notably, their synthesis is not constrained by the need for lattice matching or intricate interfacial chemistry, thus presenting a versatile foundation for the development of distinctive quantum and optoelectronic metamaterials. A Moiré pattern appears after the interlayer twist, generating a new and bigger arrangement of unitary cells. This is the source of the novel properties that characterise these systems, mostly employed to change the electronic properties by precisely controlling the twist angle as an ultimate degree of freedom [225]. Yankowitz et al. demonstrated that the electronic phase diagram of twisted 2LG could be tuned by applying an external electric field, thereby controlling the carrier density and electronic interactions within the material. This tunability is crucial for potential applications in electronic devices and quantum computing [226]. Beyond graphene, other vdW materials have also been examined under a twisted configuration. InSe presents an ultra-flat bandgap at the scale of meV at larger twist angles and some TMDs have a similar flat-band behaviour at narrower angles, also being able to achieve superconductivity in some configurations [225, 227]. Alexeev et al. reported the formation of interlayer excitons in twisted TMDs controlled by the twist angle, leading to potential applications in optoelectronics and valleytronics [228]. Mechanical characterisation of twisted 2D heterostructures is crucial for understanding their practical stability and functionality. AFM and STM are commonly employed techniques in this regard. For instance, Yoo et al. investigated the local electronic structure and strain distribution in twisted 2LG with STM [229], revealing a gradual transition from incommensurate to commensurate domains as the twist angle decreased. PFM has been employed by Li et al. to probe the electromechanical flexoelectricity in twisted double bilayer graphene (TDBG). At a small twist angle, vectorial decomposition is needed to deconvolute the small intrinsic flexoelectric response at domain walls from the background signal. In the other hand, domain walls show significant flexoelectricity due to the threefold symmetry of commensurate domains, while incommensurate areas are observed at large twist angles [230].

Finally, when several heterostructures or single nanosheets are stacked together, the three dimensions of the resulting system exceed the nanoscale. These configurations are called vertical superlattices (Fig. 2.27f) and bulk crystals (Fig. 2.27g), respectively. However, vertical superlattices maintain a certain interlayer confined quantum behaviour due to their lattice mismatch and vdW configuration [231, 232]. This vertical stack can originate from nature or via artificial synthesis, for instance, a MoS_2 on WS_2 superlattice has been formed by thermal release peeling and stacking [231]. Notable properties can be achieved in this arrangement, e.g., the change from 3D coherent to 2D incoherent charge transport in an In_2Se_3 superlattice or the modification of the electronic properties of TMDs by supporting them with a tunable h-BN superlattice [232].

Thermoelectric Phenomena

The thermoelectricity (TE) effect describes the capability of a certain material of directly producing electricity after the application of a temperature gradient (Fig. 2.28). This phenomenon is reversible, meaning that this type of elements can also be heated or cooled by a charge carrier current. This behaviour has its origin in the different principles of heat–electricity conversion, known as the Seebeck or Peltier effect for a thermal junction between two materials and as the Thomson effect for single-element TE.

The first of these TE effects was discovered in 1821 by Seebeck [184], carrying his name and describing how an electromotive force (thermoelectric voltage or thermovoltage V_{TE}) is created by proportionally heating the junction between two electrical

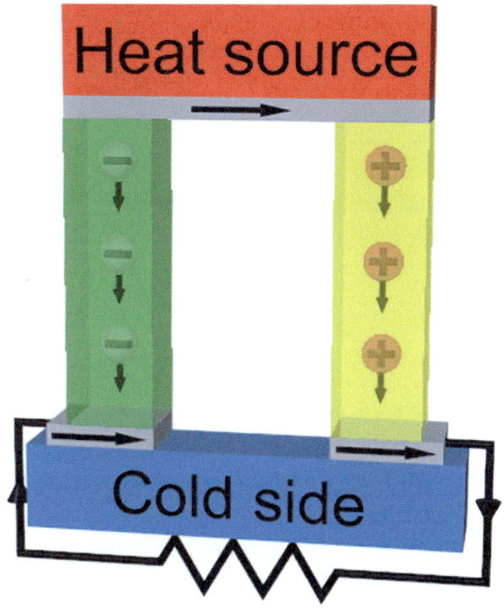

Fig. 2.28 Thermoelectric device scheme

conductors, creating a temperature difference with respect to the potential measuring points. Such proportionality is represented by the Seebeck coefficient (*S*), following the next expression:

$$S = -\frac{\Delta V}{\Delta T} = -\frac{V_{TE}}{T_h - T_c} \tag{2.15}$$

where T_h and T_c are the hot and cold temperatures, respectively. Generally, in TE semiconductors, if *S* is positive, holes dominate the charge transport (p-type), and if it is negative, electrons are the principal carriers (n-type) [184]. This electrical energy is created due to the relaxation-related diffusion of high kinetic energy electrons or holes (above the Fermi energy) in the hotter region towards the lower temperature end, knowing that the electron flow is antiparallel to the current while the hole's movement is parallel [233]. This causes an accumulation of charge carriers at the cold side, translated into negative and positive voltages at such cold spots for n-type and p-type materials, respectively, and a potential difference between the hot and cold ends [184]. Thus, the Seebeck coefficient is intricately linked to the electronic structure, which can be expressed using Mott's formula [234]:

$$S = -\frac{\pi^2 k_B^2 T}{3q}\left\{\frac{d[\ln(\sigma)]}{dW}\right\}_{W=W_F} \tag{2.16}$$

where T is the operating temperature, q is the electron charge, σ is the electrical conductivity derived with respect to the energy (W) and evaluated at the Fermi energy (W_F). This equation can be expanded substituting σ as in Eq. (2.1), providing the relations of *S* with the electronic bands' properties:

$$S = -\frac{\pi^2 k_B^2 T}{3q}\left\{\frac{dn}{ndW} + \frac{d\mu}{\mu dW}\right\}_{W=W_F} \tag{2.17}$$

where n is the carrier concentration and μ their mobility. Both depend on the carrier energy, but n can be furtherly simplified as it is equal to the product between the DOS and the Fermi function [234]. For a single parabolic band, where the energy dispersion relation is quadratic, *S* can be approximated as [234]:

$$S = \frac{8\pi^2 k_B^2 T}{3qh^2}\left(\frac{\pi}{3n}\right)^{\frac{2}{3}} m_{DOS,eff}\left(\chi + \frac{3}{2}\right) \tag{2.18}$$

with h, $m_{DOS,eff}$ and χ being the Planck's constant, the DOS effective mass and the scattering factor, respectively. The previous expressions emphasise how the DOS, the carrier mobility and the concentration contribute to *S*. Nevertheless, these formulas are derived via the Sommerfeld expansion, thus they are only valid for metals, semimetals or degenerate semiconductors. For intrinsic semiconductors, additional concepts must be considered, e.g., the doping type and the bandgap energy, resulting in different formulas to calculate the *S* [184].

The Peltier effect is the converse phenomenon, where an electric current passing through a junction of two different conductors causes a temperature difference. When current flows, electrons and holes carry not only charge but also heat. Junction carriers that transition from high to low energy states release thermal energy, while the opposite happens from low to high states, causing cooling. The Peltier coefficient is the product of the system's temperature and the S [184].

However, TE has a low conversion efficiency, given the combined requirement of high electrical and low thermal conductivities, in order to delay thermal equilibrium between the hot and cold sides [9, 235]. In nature, the thermal and electrical conductivities of materials tend to increase or decrease together since both are related to the material's electronic structure and carrier concentration, making it difficult to find a suitable TE specimen. Also, phonons contribute to such conductivity correlation by electron/phonon interactions or being scattered by elements that also influence charge carriers. In this sense, this suitability can be quantified by a formula which evaluates the performance of a TE system, called "Figure of Merit" (ZT):

$$ZT = \frac{S^2\sigma}{\kappa}T = \frac{S^2\sigma}{\kappa_e + \kappa_L}T \tag{2.19}$$

where the Seebeck parameter (S) expresses the attributes of the created electromotive force or potential difference due to the presence of an absolute thermal gradient average (T), the electrical conductivity (σ) depicts the mobility and the density of charge carriers and the thermal conductivity represents the contributions from both electrons (κ_e) and lattice vibrations (κ_L) given their shared ability of heat transfer [9, 235].

However, the "Figure of Merit" does not define the efficiency of a thermoelectric system as such. For that, additional evaluations must be employed. The first targets the operation of a thermoelectric generator, which converts heat to electrical energy via the Seebeck effect, defining its percentage efficiency (η_g) as [184]:

$$\eta_g = \frac{100(T_h - T_c)\left(\sqrt{ZT+1} - 1\right)}{T_h\left(\sqrt{ZT+1} + \frac{T_c}{T_h}\right)} \tag{2.20}$$

where T_h and T_c are the heat source and cold side (heat sink) temperatures, respectively, and ZT is the previously defined "Figure of Merit". Figure 2.29a represents the dependence of this efficiency with respect to T_h and ZT, where the efficiency clearly increases with higher values of the "Figure of Merit" as well as temperature difference. This greater heat variation enhances the Seebeck effect, generating a higher voltage (more electrical power) due to the proportional diffusion of the charge carriers from the hot to the cold side.

Reversely, to calculate the efficiency of an electricity to heat energy conversion process in a thermoelectric cooler or heater via the Peltier effect, their coefficients of performance (COP_c and COP_h, respectively) are employed. Even though being considered efficiencies since they represent the ratio of useful cooling or heating

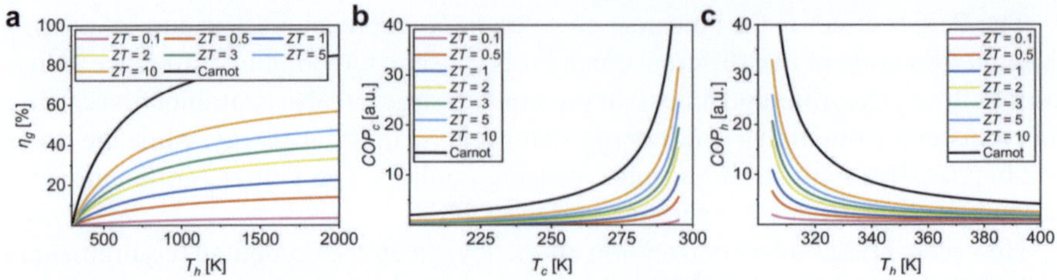

Fig. 2.29 Theoretical thermoelectric efficiencies against temperatures and "Figure of Merit". **a** Efficiency of a thermoelectric generator with fixed T_c of 300 K, **b** efficiency of a thermoelectric cooler with fixed T_h of 300 K and **c** efficiency of a thermoelectric heater with fixed T_c of 300 K

provided to electrical energy required, they are usually calculated as follows in arbitrary units instead of percentage [184, 236]:

$$COP_c = \frac{T_c\left(\sqrt{ZT + 1} - \frac{T_h}{T_c}\right)}{(T_h - T_c)\left(\sqrt{ZT + 1} + 1\right)} \tag{2.21}$$

$$COP_h = COP_c + 1 \tag{2.22}$$

By plotting such efficiency equations against their respective temperatures' variations and ZTs (Fig. 2.29b and c), one can notice a similar behaviour of increasing COP against the "Figure of Merit", validating the conventional use of the latter as a performance assessment for all the thermoelectric processes. However, the COP is observed to decrease as the temperature difference increases for both heating and cooling applications. This is due to the enhanced heat conduction through the material, which opposes the cooling or heating effect, requiring more electrical power. Additionally, higher currents needed to maintain larger temperature differences cause more Joule heating, which further reduces the efficiency [184].

It is also insightful to compare the power generation thermoelectric efficiency with other energy conversion processes widely used in today's technology (Fig. 2.30). Power production with fossil fuels is still one of the most employed methods, achieving efficiencies of 45, 52 and 42% for oil, natural gas and coal processes [237]. Interestingly, renewable energy sources like nuclear, biomass, wind and solar can compete with the efficiencies of fossil fuels, with values reported of 45% [238], 49.29% [239], 40% [240] and 47.60% [241], respectively. But hydroelectric efficiency stands well above the rest with values higher than 80% [240], proving to be the most effective energy source, while geothermal is the worst of the renewable sources, showing an efficiency of 21% [242]. With all these efficiency values, thermoelectric processes are not yet ready to compare to the aforementioned sources, given that the TE record high experimental efficiency achieved by Jiang et al. of 13.30% ($ZT = 2.70$) is still quite below the rest [243]. However, multiple studies are

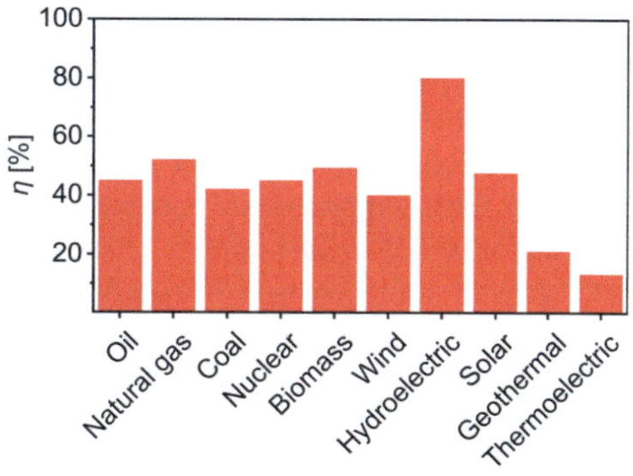

Fig. 2.30 Maximum achieved efficiencies of energy conversion processes

performed in the current research due to its potential and simple performance mechanism involving no moving parts, little maintenance and no pollution emissions, with many platforms for efficiency enhancement.

Certain nanomaterials can be engineered to increase this efficiency. For instance, vdW structures have become valid candidates in said field, serving its ultimate purpose given their tunability and allowing the optimisation of S, σ and κ to finally enhance the "Figure of Merit".

Graphene possesses a large phononic mean free path that translates into its elevated thermal conductivity, being counterproductive in TE systems. Thus, researchers focus on reducing its heat transport capabilities by increasing the number of layers to benefit from the resulting scattering of the cross-plane acoustic modes, approaching the κ values of graphite [54]. Another strategy is the employment of a SiO_2 substrate, which also reduces the thermal conductivity by a factor of 5. The electrical influence in the "Figure of Merit" formula is represented in its numerator by the product of the squared Seebeck coefficient and the electrical conductivity, frequently simplifying it as the power factor ($S^2\sigma$). In such regard, the electronic properties of graphene become a valuable resource to modulate the TE yield. Studies have shown that graphene diverges from the well-accepted Mott relation for metallic samples, which predicted low S in semimetallic graphene given its direct proportionality with the asymmetry of the DOS around the Fermi level. However, the studies observed anomalous Seebeck coefficients close to the CNP, also known as Dirac point, validating it via experimental research of cleaner graphene samples on h-BN that verified the electron/electron scattering domination at high temperatures in the hydrodynamic regime [54].

A theoretical study performed on TMDs predicted their superior ZT values after tuning their electronic band structure via the addition or removal of monolayers or by inducing strain in the in-plane direction [88]. Experimental investigation corroborates this proper thermoelectric efficiency, e.g., randomly aligned WS_2 was found to have a record low thermal conductivity [244] and doped MoS_2 showed an elevated power factor (8.50 m W m^{-1} K^{-2}) at RT [62]. The boosted S was attributed to the confinement-dependent large effective mass and valley degeneracy obtained as the Fermi level approached the bottom of the conduction band.

Besides the use of vdW materials, charge carrier energy filtering is another promising strategy to enhance the TE performance. This approach involves selectively scattering charge carriers based on their energy, allowing high energy electrons or holes to contribute more effectively to electrical conduction while blocking lower energy charges with potential barriers. Also, such barriers suppress bipolar effects by obstructing the flow of minority charge carriers, preventing a significant reduction in electrical conductivity. Ultimately, the filtering mechanism aims to improve the power factor by effectively increasing the average energy of the carriers [245].

To understand the physical base of the carrier filtering effect, the Boltzmann transport equation must be solved. This expression represents a balance between external fields (e.g., temperature gradients or electrical fields) and scattering events for slight deviations from equilibrium using the relaxation time approximation. Thus, its derivation allows expressing σ and S as functions of the energy-dependent relaxation times of charge carriers (ζ), DOS (n_{DOS}) and Fermi–Dirac equilibrium distribution function ($\bar{n}$) as well as their effective mass (m_{eff}), and the Fermi energy [245].

$$\sigma = -\frac{2q^2 W \zeta n_{DOS}}{3m_{eff}} \frac{\partial \bar{n}}{\partial W} \tag{2.23}$$

$$S = -\frac{3m_{eff}}{2q^3 T} \langle W - W_F \rangle_\sigma \tag{2.24}$$

In semiconductors, an asymmetric DOS with respect to the Fermi level yields large S values. Hence, modifying the charge carrier relaxation time to extend the level of asymmetry of its product with the DOS can further enhance S, thus establishing the EF effect. Also, while high mobility is generally desirable, the presence of energy barriers can scatter high energy carriers if not carefully designed. However, theoretical models suggest that the optimal barrier height can maximise the energy filtering effect without severely impacting mobility [246].

It is clear that achieving significant power factor improvement requires materials with controlled, stable, and optimised layered structures. Reports indicate that nanograined, nanoporous, nanocomposite materials, and bulk materials with core/shell heterostructures enable selective scattering of low energy charge carriers by potential barriers [245]. Lastly, in such filtering phenomenon, κ remains relatively unaffected, even being reduced due to increased phonon scattering at the nanostructures interfaces, therefore a significant improvement in ZT can be achieved [246].

Electromechanical Two-Dimensional Devices

By definition, an electromechanical device is a structure capable of converting a mechanical input to an electronic output and vice versa.

Classic examples in this area are the generator, which uses a mechanical process to produce electricity, and the motor, which uses electrical power to output a certain mechanical response [248]. With the current trend of downsizing the components in devices, 2D materials show promise as elementary structures to build a successful electromechanical system. Piezoelectricity generally dominates the materials' properties in this regard, giving them the ability to transform to electricity different types of mechanical stress or trigger an expansion or contraction in the system after injecting a charge current. In vdW materials, the origin of this property is attributed to the generation of polarised charges in response to an external mechanical stimulus due to their non-symmetric structure (Fig. 2.31) [247].

Triboelectricity is another electromechanical phenomenon based on the contact electrification effect, which explains how two different electrically charged materials achieve equilibrium after they come into contact, causing an electrostatic potential that drives the induced electrons in the outer circuit loop [249].

Interesting piezo- and triboelectric applications have been developed with 2D devices [250]. In the piezoelectric field, black phosphorus nanosheets have been used to fabricate a mechano-electrochemical device capable of harvesting low frequency mechanical energy. Also, monolayer WSe_2 has been employed as a building block for a piezoelectric nanogenerator (PENG). Similarly, Kim et al. assembled a graphene-based flexible and transparent triboelectric nanogenerator (TENG) capable of powering commercial light-emitting diodes (LEDs) from electrical signals originated by a mechanical source.

Quantifying the relation between the piezoelectric displacement and the applied electrical field is crucial to fully comprehend the electromechanical phenomenon. The following equation is used in devices measured by AFM technologies [251]:

$$A_p = \alpha d_{ij} V_{AC} \tag{2.25}$$

where A_p is the piezoelectric amplitude, α is a calibration parameter determined from the LIA and the optical sensitivity, d_{ij} is the locally relevant element of the third-rank d_{ijk} piezoelectric tensor and V_{AC} is the applied voltage.

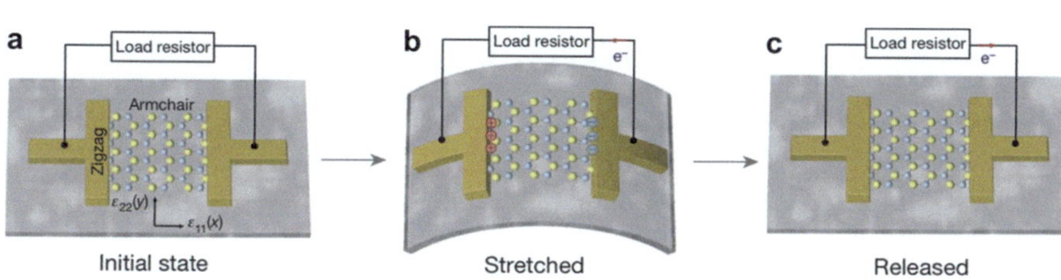

Fig. 2.31 Piezoelectric monolayer MoS_2 device. **a** Initial resting device, **b** stretched state where electrical charges are generated at the Z-Z edges of the monolayer and **c** released stretched device returns to the initial state and liberates a current flow. Adapted with permission from Ref. [247], Copyright © 2014 Springer Nature [247]

2.6 Summary

Layered materials are undoubtedly a pathway to enhance the performance of multidisciplinary developments in branches ranging from medicine to energy conversion. This review provides a condensed analysis about 2D materials with the purpose of conveying the fundamental knowledge essential to comprehend the following chapters. Specifically, the nature of four different vdW structures is initially discussed, yielding an insight into their physical properties. Also, prevalent and recently developed 2D materials' synthesis and characterisation techniques are addressed herein, followed by a section that elucidates the advanced thermal behaviour and the exotic multiphysical interactions that these structures can present. The interplay of this advanced physical phenomena is currently the research focus in the field, with notorious steps forward in the related superconductivity, anisotropy or thermoelectricity. Hence, the last segment of the review can be understood as an outline of the subsequent sections, where these phenomena are expanded and further investigated.

References

1. Velický, M., & Toth, P. S. (2017). From two-dimensional materials to their heterostructures: An electrochemist's perspective. *Applied Materials Today, 8*, 68–103.
2. Geim, A. K., & Novoselov, K. S. (2007). The rise of graphene. *Nature Materials, 6*, 183–191.
3. Wallace, P. R. (1947). The band theory of graphite. *Physical Review, 71*, 622–634.
4. Novoselov, K. S., et al. (2004). Electric field effect in atomically thin carbon films. *Science, 306*, 666–669.
5. Novoselov, K. S., Mishchenko, A., Carvalho, A. & Castro Neto, A. H. (2016). 2D materials and van der Waals heterostructures. *Science, 353*, aac9439.
6. ISO. (2017). *ISO/TS 80004–13:2017(en) nanotechnologies—Vocabulary—Part 13: graphene and related two-dimensional (2D) materials.* https://www.iso.org/obp/ui/#iso:std:iso:ts:80004:-13:ed-1:v1:en.
7. Wang, J., Wang, L., Xing, Y. H., Liu, T. X., & Zhang, S. T. (2018). Mechanism of large tunable thermal transport in graphene with oxygen functional groups. *Journal of Applied Physics, 124*, 175108.
8. Dubois, S. M. M., Zanolli, Z., Declerck, X., & Charlier, J. C. (2009). Electronic properties and quantum transport in graphene-based nanostructures. *European Physical Journal B: Condensed Matter and Complex Systems, 72*, 1–24.
9. Wu, J., Chen, Y. B., Wu, J. Q., & Hippalgaonkar, K. (2018). Perspectives on thermoelectricity in layered and 2D materials. *Advanced Electronic Materials, 4*, 1800248.
10. Cooper, D. R., et al. (2012). Experimental review of graphene. *ISRN Condensed Matter Physics, 2012*, 1–56.
11. Parvez, K. (2019). Two-dimensional nanomaterials: crystal structure and synthesis. In M. Nurunnabi & J. R. McCarthy (Eds.), *Biomedical applications of graphene and 2D nanomaterials* (Chap. 1, pp. 1–25). Elsevier.
12. Xu, M., Liang, T., Shi, M., & Chen, H. (2013). Graphene-like two-dimensional materials. *Chemical Reviews, 113*, 3766–3798.
13. Mak, K. F., Lee, C., Hone, J., Shan, J., & Heinz, T. F. (2010). Atomically thin MoS_2: A new direct-gap semiconductor. *Physical Review Letters, 105*, 136805.
14. Xia, J., Yan, J., & Shen, Z. X. (2017). Transition metal dichalcogenides: Structural, optical and electronic property tuning via thickness and stacking. *FlatChem, 4*, 1–19.

15. Feng, W., et al. (2015). Synthesis, properties and applications of 2D non-graphene materials. *Nanotechnology, 26*, 292001.
16. Lin, J., et al. (2018). Indium selenide monolayer: A two-dimensional material with strong second harmonic generation. *CrystEngComm, 20*, 2573–2582.
17. Sun, Y., et al. (2018). InSe: A two-dimensional material with strong interlayer coupling. *Nanoscale, 10*, 7991–7998.
18. Sang, D. K., et al. (2019). Two dimensional β-InSe with layer-dependent properties: Band alignment, work function and optical properties. *Nanomaterials, 9*, 82.
19. Buckley, D., et al. (2021). Anomalous low thermal conductivity of atomically thin InSe probed by scanning thermal microscopy. *Advanced Functional Materials, 31*, 2008967.
20. Mafi, E., Soudi, A., & Gu, Y. (2012). Electronically driven amorphization in phase-change In_2Se_3 nanowires. *Journal of Physical Chemistry C, 116*, 22539–22544.
21. Sun, Y., et al. (2020). New polymorphs of 2D indium selenide with enhanced electronic properties. *Advanced Functional Materials, 30*, 2001920.
22. Lin, Y., & Connell, J. W. (2012). Advances in 2D boron nitride nanostructures: Nanosheets, nanoribbons, nanomeshes, and hybrids with graphene. *Nanoscale, 4*, 6908–6939.
23. Wang, Z., et al. (2020). Dispersion of high-quality boron nitride nanosheets in polyethylene for nanocomposites of superior thermal transport properties. *Nanoscale Advances, 2*, 2507–2513.
24. Golberg, D., et al. (2010). Boron nitride nanotubes and nanosheets. *ACS Nano, 4*, 2979–2993.
25. Xu, X., Chen, J., & Li, B. (2016). Phonon thermal conduction in novel 2D materials. *Journal of Physics: Condensed Matter, 28*, 483001.
26. Schiefele, J., Sols, F., & Guinea, F. (2012). Temperature dependence of the conductivity of graphene on boron nitride. *Physical Review B, 85*, 195420.
27. Sachs, B., Wehling, T. O., Katsnelson, M. I., & Lichtenstein, A. I. (2011). Adhesion and electronic structure of graphene on hexagonal boron nitride substrates. *Physical Review B, 84*, 195414.
28. Noor-A-Alam, M., Kim, H. J., & Shin, Y.-H. (2014). Dipolar polarization and piezoelectricity of a hexagonal boron nitride sheet decorated with hydrogen and fluorine. *Physical Chemistry Chemical Physics: PCCP, 16*, 6575–6582.
29. Naik, S. G., & Rabinal, M. H. K. (2020). Molybdenum disulphide heterointerfaces as potential materials for solar cells, energy storage, and hydrogen evolution. *Energy Technology, 8*, 1901299.
30. Wu, D., et al. (2015). Thickness-dependent dielectric constant of few-layer In_2Se_3 nanoflakes. *Nano Letters, 15*, 8136–8140.
31. Xue, F., et al. (2018). Room-temperature ferroelectricity in hexagonally layered α-In_2Se_3 nanoflakes down to the monolayer limit. *Advanced Functional Materials, 28*, 1803738.
32. Golla, D., et al. (2013). Optical thickness determination of hexagonal boron nitride flakes. *Applied Physics Letters, 102*, 161906.
33. Liu, M., Artyukhov, V. I., Lee, H., Xu, F., & Yakobson, B. I. (2013). Carbyne from first principles: Chain of C atoms, a nanorod or a nanorope. *ACS Nano, 7*, 10075–10082.
34. Cao, K., et al. (2020). Elastic straining of free-standing monolayer graphene. *Nature Communications, 11*, 1–7.
35. Clark, N., Oikonomou, A., & Vijayaraghavan, A. (2013). Ultrafast quantitative nanomechanical mapping of suspended graphene. *Physica Status Solidi B, 250*, 2672–2677.
36. Wong, C. L., Annamalai, M., Wang, Z. Q., & Palaniapan, M. (2010). Characterization of nanomechanical graphene drum structures. *Journal of Micromechanics and Microengineering, 20*, 115029.
37. Ni, Z. H., et al. (2009). Anisotropic mechanical properties of graphene sheets from molecular dynamics. *Physics B, 405*, 1301–1306.
38. Tu, Q., et al. (2017). Interfacial mechanical properties of graphene on self-assembled monolayers: Experiments and simulations. *ACS Applied Materials & Interfaces, 9*, 10203–10213.
39. Hajgato, B., et al. (2013). Out-of-plane shear and out-of plane Young's modulus of double-layer graphene. *Chemical Physics Letters, 564*, 37–40.

40. Bertolazzi, S., Brivio, J., & Kis, A. (2011). Stretching and breaking of ultrathin MoS_2. *ACS Nano, 5*, 9703–9709.
41. Li, Y., et al. (2018). Mapping the elastic properties of two-dimensional MoS_2 via bimodal atomic force microscopy and finite element simulation. *npj Computational Materials, 4*, 49.
42. Castellanos-Gomez, A., et al. (2012). Mechanical properties of freely suspended semiconducting graphene-like layers based on MoS_2. *Nanoscale Research Letters, 7*, 1–4.
43. Castellanos-Gomez, A., et al. (2012). Elastic properties of freely suspended MoS_2 nanosheets. *Advanced Materials, 24*, 772–775.
44. Deng, S., Li, L., & Li, M. (2018). Stability of direct band gap under mechanical strains for monolayer MoS_2, $MoSe_2$, WS_2 and WSe_2. *Physics E, 101*, 44–49.
45. Li, J., Medhekar, N. V., & Shenoy, V. B. (2013). Bonding charge density and ultimate strength of monolayer transition metal dichalcogenides. *Journal of Physical Chemistry C, 117*, 15842–15848.
46. Raranskii, N. D., Balazyuk, V. N., Kovalyuk, Z. D., Mel'Nik, N. I., & Gevik, V. B. (2011). Crystal growth and elastic properties of In_2Se_3. *Inorganic Materials, 47*, 1174–1177.
47. Falin, A., et al. (2017). Mechanical properties of atomically thin boron nitride and the role of interlayer interactions. *Nature Communications, 8*, 1–9.
48. Mirnezhad, M., Ansari, R., & Rouhi, H. (2013). Mechanical properties of multilayer boron nitride with different stacking orders. *Superlattices and Microstructures, 53*, 223–231.
49. Wu, J., Wang, B., Wei, Y., Yang, R., & Dresselhaus, M. (2013). Mechanics and mechanically tunable band gap in single-layer hexagonal boron-nitride. *Materials Research Letters, 1*, 200–206.
50. Lee, C., Wei, X., Kysar, J. W., & Hone, J. (2008). Measurement of the elastic properties and intrinsic strength of monolayer graphene. *Science, 321*, 385–388.
51. Zhao, H., Min, K., & Aluru, N. R. (2009). Size and chirality dependent elastic properties of graphene nanoribbons under uniaxial tension. *Nano Letters, 9*, 3012–3015.
52. Wong, C. H., & Vijayaraghavan, V. (2012). Nanomechanics of free form and water submerged single layer graphene sheet under axial tension by using molecular dynamics simulation. *Materials Science and Engineering A, 556*, 420–428.
53. Hossain, M. Z., et al. (2018). Anisotropic toughness and strength in graphene and its atomistic origin. *Journal of the Mechanics and Physics of Solids, 110*, 118–136.
54. Amollo, T. A., Mola, G. T., Kirui, M. S. K., & Nyamori, V. O. (2018). Graphene for thermoelectric applications: Prospects and challenges. *Critical Reviews in Solid State and Materials Sciences, 43*, 133–157.
55. Cao, M., et al. (2019). Ultrahigh electrical conductivity of graphene embedded in metals. *Advanced Functional Materials, 29*, 1806792.
56. Chen, J., & Gao, X. (2019). Directional dependence of electrical and thermal properties in graphene-nanoplatelet-based composite materials. *Results in Physics, 15*, 102608.
57. Wang, H., et al. (2016). Simultaneous measurement of electrical and thermal conductivities of suspended monolayer graphene. *Journal of Applied Physics, 119*, 244306.
58. Murata, H., et al. (2019). High-electrical-conductivity multilayer graphene formed by layer exchange with controlled thickness and interlayer. *Science and Reports, 9*, 4068.
59. Shayeganfar, F. (2022). Strain engineering of electronic properties and anomalous valley hall conductivity of transition metal dichalcogenide nanoribbons. *Science and Reports, 12*, 11285.
60. Zhao, P., et al. (2020). Electronic and optical properties of transition metal dichalcogenides under symmetric and asymmetric field-effect doping. *New Journal of Physics, 22*, 083072.
61. Chen, R. S., Tang, C. C., Shen, W. C., & Huang, Y. S. (2014). Thickness-dependent electrical conductivities and ohmic contacts in transition metal dichalcogenides multilayers. *Nanotechnology, 25*, 415706.
62. Hippalgaonkar, K., et al. (2017). High thermoelectric power factor in two-dimensional crystals of MoS_2. *Physical Review B, 95*, 115407.
63. Schmidt, H., Giustiniano, F., & Eda, G. (2015). Electronic transport properties of transition metal dichalcogenide field-effect devices: Surface and interface effects. *Chemical Society Reviews, 44*, 7715–7736.

64. Kim, N. O., et al. (2001). Electrical and optical properties of α-In_2Se_3 single crystals with an indium excess. *Journal of the Korean Physical Society, 38*, 405–408.
65. Emziane, M., Marsillac, S., & Bernède, J. C. (2000). Preparation of highly oriented α-In_2Se_3 thin films by a simple technique. *Materials Chemistry and Physics, 62*, 84–87.
66. Wang, J., Ma, F., Liang, W., & Sun, M. (2017). Electrical properties and applications of graphene, hexagonal boron nitride (h-BN), and graphene/h-BN heterostructures. *Materials Today Physics, 2*, 6–34.
67. Steinborn, C., et al. (2013). Correlation between microstructure and electrical resistivity of hexagonal boron nitride ceramics. *Journal of the European Ceramic Society, 33*, 1225–1235.
68. Uddin, M. R., et al. (2014). Electrical transport properties of (BN)-rich hexagonal (BN)C semiconductor alloys. *AIP Advances, 4*, 087141.
69. Gunst, T., Markussen, T., Stokbro, K., & Brandbyge, M. (2016). First-principles method for electron-phonon coupling and electron mobility: Applications to two-dimensional materials. *Physical Review B, 93*, 035414.
70. Poncé, S., Li, W., Reichardt, S., & Giustino, F. (2020). First-principles calculations of charge carrier mobility and conductivity in bulk semiconductors and two-dimensional materials. *Reports on Progress in Physics, 83*, 036501.
71. Novoselov, K. S., et al. (2005). Two-dimensional gas of massless Dirac fermions in graphene. *Nature, 438*, 197–200.
72. Norimatsu, W. (2023). A review on carrier mobilities of epitaxial graphene on silicon carbide. *Materials, 16*, 7668.
73. Armakavicius, N., et al. (2021). Resolving mobility anisotropy in quasi-free-standing epitaxial graphene by terahertz optical Hall effect. *Carbon, 172*, 248–259.
74. Cheng, L., & Liu, Y. (2018). What limits the intrinsic mobility of electrons and holes in two dimensional metal dichalcogenides? *Journal of the American Chemical Society, 140*, 17895–17900.
75. Dagan, R., Vaknin, Y., Weisman, D., Amit, I., & Rosenwaks, Y. (2019). Accurate method to determine the mobility of transition-metal dichalcogenides with incomplete gate screening. *ACS Applied Materials & Interfaces, 11*, 44406–44412.
76. Rawat, A., Jena, N., Dimple, & De Sarkar, A. (2018). A comprehensive study on carrier mobility and artificial photosynthetic properties in group VI B transition metal dichalcogenide monolayers. *Journal of Materials Chemistry, A**6***, 8693–8704.
77. Bandurin, D. A., et al. (2017). High electron mobility, quantum Hall effect and anomalous optical response in atomically thin InSe. *Nature Nanotechnology, 12*, 223–227.
78. Li, M., et al. (2018). High mobilities in layered InSe transistors with indium-encapsulation-induced surface charge doping. *Advanced Materials, 30*, 1803690.
79. Khatami, M. M., Van de Put, M. L., & Vandenberghe, W. G. (2021). First-principles study of electronic transport in germanane and hexagonal boron nitride. *Physical Review B, 104*, 235424.
80. Doan, T. C., Li, J., Lin, J. Y., & Jiang, H. X. (2014). Charge carrier transport properties in layer structured hexagonal boron nitride. *AIP Advances, 4*, 107126.
81. Lee, J. U., Yoon, D., Kim, H., Lee, S. W., & Cheong, H. (2011). Thermal conductivity of suspended pristine graphene measured by Raman spectroscopy. *Physical Review B, 83*, 081419.
82. Balandin, A. A., et al. (2008). Superior thermal conductivity of single-layer graphene. *Nano Letters, 8*, 902–907.
83. Lindsay, L., Broido, D. A., & Mingo, N. (2011). Flexural phonons and thermal transport in multilayer graphene and graphite. *Physical Review B, 83*, 235428.
84. Cui, L., Shi, S. Q., Wei, G. S., & Du, X. Z. (2018). Shear deformation-induced anisotropic thermal conductivity of graphene. *Physical Chemistry Chemical Physics: PCCP, 20*, 951–957.
85. Pop, E., Varshney, V., & Roy, A. K. (2012). Thermal properties of graphene: Fundamentals and applications. *MRS Bulletin, 37*, 1273–1281.
86. Sahoo, S., Gaur, A. P. S., Ahmadi, M., Guinel, M. J. F., & Katiyar, R. S. (2013). Temperature-dependent Raman studies and thermal conductivity of few-layer MoS_2. *Journal of Physical Chemistry C, 117*, 9042–9047.

87. Zhang, Z. W., Xie, Y., Ouyang, Y. L., & Chen, Y. P. (2017). A systematic investigation of thermal conductivities of transition metal dichalcogenides. *International Journal of Heat and Mass Transfer, 108*, 417–422.
88. Pallecchi, I., Manca, N., Patil, B., Pellegrino, L., & Marre, D. (2020). Review on thermoelectric properties of transition metal dichalcogenides. *Nano Futures, 4*, 032008.
89. Liu, J., Choi, G.-M., & Cahill, D. G. (2014). Measurement of the anisotropic thermal conductivity of molybdenum disulfide by the time-resolved magneto-optic Kerr effect. *Journal of Applied Physics, 116*, 233107.
90. Muratore, C., et al. (2014). Thermal anisotropy in nano-crystalline MoS_2 thin films. *Physical Chemistry Chemical Physics: PCCP, 16*, 1008–1014.
91. Zhou, S., Tao, X., & Gu, Y. (2016). Thickness-dependent thermal conductivity of suspended two-dimensional single-crystal In_2Se_3 layers grown by chemical vapor deposition. *Journal of Physical Chemistry C, 120*, 4753–4758.
92. Rai, A., Sangwan, V. K., Gish, J. T., Hersam, M. C., & Cahill, D. G. (2021). Anisotropic thermal conductivity of layered indium selenide. *Applied Physics Letters, 118*, 073101.
93. Hsin, C.-L., Huang, J.-H., Spiewak, P., Ciupiński, Ł, & Lee, S.-W. (2019). Anisotropy of thermal conductivity in In_2Se_3 nanostructures. *Applied Surface Science, 494*, 867–870.
94. Wang, C., et al. (2016). Superior thermal conductivity in suspended bilayer hexagonal boron nitride. *Science and Reports, 6*, 25334.
95. Jaffe, G. R., et al. (2023). Thickness-dependent cross-plane thermal conductivity measurements of exfoliated hexagonal boron nitride. *ACS Applied Materials & Interfaces, 15*, 12545–12550.
96. Sood, A., et al. (2021). Engineering thermal transport across layered graphene-MoS_2 superlattices. *ACS Nano, 15*, 19503–19512.
97. Liu, Y., et al. (2017). Thermal conductance of the 2D MoS_2/h-BN and graphene/h-BN interfaces. *Science and Reports, 7*, 43886.
98. Mak, K. F., Lui, C. H., & Heinz, T. F. (2010). Measurement of the thermal conductance of the graphene/SiO_2 interface. *Applied Physics Letters, 97*, 221904.
99. Yue, J., et al. (2024). Unraveling the mechanisms of thermal boundary conductance at the graphene-silicon interface: Insights from ballistic, diffusive, and localized phonon transport regimes. *Physical Review B, 109*, 115302.
100. Zhang, X., et al. (2015). Measurement of lateral and interfacial thermal conductivity of single- and bilayer MoS_2 and $MoSe_2$ using refined optothermal Raman technique. *ACS Applied Materials & Interfaces, 7*, 25923–25929.
101. Yalon, E., et al. (2017). Energy dissipation in monolayer MoS_2 electronics. *Nano Letters, 17*, 3429–3433.
102. Li, X., et al. (2017). Thermal conduction across a boron nitride and SiO_2 interface. *Journal of Physics D: Applied Physics, 50*, 104002.
103. Wang, X.-Y., Narita, A., & Müllen, K. (2017). Precision synthesis versus bulk-scale fabrication of graphenes. *Nature Reviews Chemistry, 2*, 0100.
104. Liang, X., et al. (2009). Electrostatic force assisted exfoliation of prepatterned few-layer graphenes into device sites. *Nano Letters, 9*, 467–472.
105. Moldt, T., et al. (2011). High-yield production and transfer of graphene flakes obtained by anodic bonding. *ACS Nano, 5*, 7700–7706.
106. Bottcher, A., et al. (2012). Nanostructured arrays of stacked graphene sheets. *Nanotechnology, 23*, 415302.
107. Shen, J., et al. (2015). Liquid phase exfoliation of two-dimensional materials by directly probing and matching surface tension components. *Nano Letters, 15*, 5449–5454.
108. Zheng, W., & Lee, L. Y. S. (2022). Beyond sonication: Advanced exfoliation methods for scalable production of 2D materials. *Matter, 5*, 515–545.
109. Huang, Y., et al. (2015). Reliable exfoliation of large-area high-quality flakes of graphene and other two-dimensional materials. *ACS Nano, 9*, 10612–10620.
110. Ferrari, A. C., et al. (2015). Science and technology roadmap for graphene, related two-dimensional crystals, and hybrid systems. *Nanoscale, 7*, 4598–4810.

111. Leong, W. S., et al. (2019). Paraffin-enabled graphene transfer. *Nature Communications, 10*, 867.
112. Lotsch, B. V. (2015). Vertical 2D heterostructures. *Annual Review of Materials Research, 45*, 85–109.
113. Sun, L., et al. (2021). Chemical vapour deposition. *Nature Reviews Methods Primers, 1*, 1–20.
114. Kwon, H., et al. (2016). Large scale MoS_2 nanosheet logic circuits integrated by photolithography on glass. *2D Materials, 3*, 044001.
115. Okazaki, S. (2015). High resolution optical lithography or high throughput electron beam lithography: The technical struggle from the micro to the nano-fabrication evolution. *Microelectronic Engineering, 133*, 23–35.
116. Ito, T., & Okazaki, S. (2000). Pushing the limits of lithography. *Nature, 406*, 1027–1031.
117. Manfrinato, V. R., et al. (2019). Patterning Si at the 1 nm length scale with aberration-corrected electron-beam lithography: Tuning of plasmonic properties by design. *Advanced Functional Materials, 29*, 1903429.
118. Penczek, P. A. (2010). Resolution measures in molecular electron microscopy. In G. J. Jensen (Ed.), *Methods in enzymology* (Vol. 482, Chap. 3, pp. 73–100). Academic Press.
119. Terris, B. D., & Thomson, T. (2005). Nanofabricated and self-assembled magnetic structures as data storage media. *Journal of Physics D: Applied Physics, 38*, R199–R222.
120. Liming, R., & Baoqin, C. (2004). In *International conference on solid-state and integrated circuits technology* (7th ed., pp. 579–582). IEEE.
121. Heusinger, M., Banasch, M., & Zeitner, U. D. (2017). Rowland ghost suppression in high efficiency spectrometer gratings fabricated by e-beam lithography. *Optics Express, 25*, 6182–6191.
122. EESemi. (2004). *Electron beam lithography*. https://www.eesemi.com/lith_electron.htm
123. Vignal, V., et al. (2005). Mapping the 3D-surface strain field of patterned tensile stainless steels using atomic force microscopy. *Ultramicroscopy, 103*, 183–189.
124. Simbulan, K. B. C., Chen, P.-C., Lin, Y.-Y., & Lan, Y.-W. (2018). A standard and reliable method to fabricate two-dimensional nanoelectronics. *Journal of Visualized Experiments, 138*, 57885.
125. Murakami, K., Kadowaki, T., & Fujita, J. I. (2013). Damage and strain in single-layer graphene induced by very-low-energy electron-beam irradiation. *Applied Physics Letters, 102*, 043111.
126. Liu, X., et al. (2020). Thermomechanical nanostraining of two-dimensional materials. *Nano Letters, 20*, 8250–8257.
127. Howell, S. T., Grushina, A., Holzner, F., & Brugger, J. (2020). Thermal scanning probe lithography—A review. *Microsystems & Nanoengineering, 6*, 21.
128. Garcia, R., Knoll, A. W., & Riedo, E. (2014). Advanced scanning probe lithography. *Nature Nanotechnology, 9*, 577–587.
129. Tseng, A. A., Notargiacomo, A., & Chen, T. P. (2005). Nanofabrication by scanning probe microscope lithography: a review. *Journal of Vacuum Science & Technology B: Microelectronics and Nanometer Structures—Processing, Measurement, Phenomena, 23*, 877–894.
130. Rank, R., Brückl, H., Kretz, J., Mönch, I., & Reiss, G. (1997). Nanoscale modification of conducting lines with a scanning force microscope. *Vacuum, 48*, 467–472.
131. Filho, H. D. F., Maurício, M. H. P., Ponciano, C. R., & Prioli, R. (2004). Metal layer mask patterning by force microscopy lithography. *Materials Science and Engineering B, 112*, 194–199.
132. Fenwick, O., et al. (2009). Thermochemical nanopatterning of organic semiconductors. *Nature Nanotechnology, 4*, 664–668.
133. Allain, A., Kang, J. H., Banerjee, K., & Kis, A. (2015). Electrical contacts to two-dimensional semiconductors. *Nature Materials, 14*, 1195–1205.
134. Wang, S., Yu, Z., & Wang, X. (2018). Electrical contacts to two-dimensional transition-metal dichalcogenides. *Journal of Semiconductors, 39*, 124001.
135. Leonard, F., & Talin, A. A. (2011). Electrical contacts to one- and two-dimensional nanomaterials. *Nature Nanotechnology, 6*, 773–783.

136. Bennig, G. K. (1986). Atomic force microscope and method for imaging surfaces with atomic resolution. United States patent 4724318.
137. Zhang, H., et al. (2018). Atomic force microscopy for two-dimensional materials: A tutorial review. *Optics Communication, 406*, 3–17.
138. Umeda, K., McArthur, S. J., & Kodera, N. (2023). Spatiotemporal resolution in high-speed atomic force microscopy for studying biological macromolecules in action. *Microscopy, 72*, 151–161.
139. Crewe, A. V. (1963). Scanning electron microscope. United States patent 3191028.
140. Egerton, R. F. (2016). *Physical principles of electron microscopy* (2nd ed.). Springer Cham.
141. Li, R., et al. (2018). Optical characterizations of two-dimensional materials using nonlinear optical microscopies of CARS, TPEF, and SHG. *Nanophotonics, 7*, 873–881.
142. Terpiłowski, K., & Szaniawska, M. (2019). Surface properties of graphene and graphene/diamond composites located at a substrate with tungsten carbide doped by metals composites. *Adsorption, 25*, 513–520.
143. Yin, P., Lin, Q., & Duan, Y. (2020). Applications of Raman spectroscopy in two-dimensional materials. *ournal of Innovative Optical Health Sciences, 13*, 2030010.
144. Cong, X., Liu, X.-L., Lin, M.-L., & Tan, P.-H. (2020). Application of Raman spectroscopy to probe fundamental properties of two-dimensional materials. *npj 2D Materials and Applications, 4*, 1–12.
145. Raman, C. V., & Krishnan, K. S. (1928). A new type of secondary radiation. *Nature, 121*, 501–502.
146. Shao, F., et al. (2018). Chemical mapping of nanodefects within 2D covalent monolayers by tip-enhanced raman spectroscopy. *ACS Nano, 12*, 5021–5029.
147. Greczynski, G., & Hultman, L. (2020). X-ray photoelectron spectroscopy: Towards reliable binding energy referencing. *Progress in Materials Science, 107*, 100591.
148. Stevie, F. A., & Donley, C. L. (2020). Introduction to X-ray photoelectron spectroscopy. *Journal of Vacuum Science & Technology A, 38*, 063204.
149. Dazzi, A., & Prater, C. B. (2017). AFM-IR: Technology and applications in nanoscale infrared spectroscopy and chemical imaging. *Chemical Reviews, 117*, 5146–5173.
150. Jalili, N., & Laxminarayana, K. (2004). A review of atomic force microscopy imaging systems: Application to molecular metrology and biological sciences. *Mechatronics, 14*, 907–945.
151. Garcia, R., & Perez, R. (2002). Dynamic atomic force microscopy methods. *Surface Science Reports, 47*, 197–301.
152. Soergel, E. (2011). Piezoresponse force microscopy (PFM). *Journal of Physics D: Applied Physics, 44*, 464003.
153. Jesse, S., Baddorf, A. P., & Kalinin, S. V. (2006). Dynamic behaviour in piezoresponse force microscopy. *Nanotechnology, 17*, 1615–1628.
154. Jesse, S., et al. (2010). Resolution theory, and static and frequency-dependent cross-talk in piezoresponse force microscopy. *Nanotechnology, 21*, 405703.
155. Kim, S., Seol, D., Lu, X., Alexe, M., & Kim, Y. (2017). Electrostatic-free piezoresponse force microscopy. *Scientific Reports, 7*, 41657.
156. Seidel, J., et al. (2009). Conduction at domain walls in oxide multiferroics. *Nature Materials, 8*, 229–234.
157. Kolosov, O., Gruverman, A., Hatano, J., Takahashi, K., & Tokumoto, H. (1995). Nanoscale visualization and control of ferroelectric domains by atomic force microscopy. *Physical Review Letters, 74*, 4309–4312.
158. Yamanaka, K., Ogiso, H., & Kolosov, O. (1994). Ultrasonic force microscopy for nanometer resolution subsurface imaging. *Applied Physics Letters, 64*, 178–180.
159. Dinelli, F., Briggs, G., Kolosov, O., & Biswas, S. (2000). Measurements of stiff-material compliance on the nanoscale using ultrasonic force microscopy. *Physical Review B, 61*, 13995–14006.
160. Tinker-Mill, C., Mayes, J., Allsop, D., & Kolosov, O. (2014). Ultrasonic force microscopy for nanomechanical characterization of early and late-stage amyloid-ß peptide aggregation. *Science and Reports, 4*, 1–7.

161. Dinelli, F., Albonetti, C., & Kolosov, O. V. (2011). Ultrasonic force microscopy: Detection and imaging of ultra-thin molecular domains. *Ultramicroscopy, 111*, 267–272.
162. Skilbeck, M. S., et al. (2014). Multimodal microscopy using 'half and half' contact mode and ultrasonic force microscopy. *Nanotechnology, 25*, 335708.
163. Robinson, B. J., et al. (2015). Structural, optical and electrostatic properties of single and few-layers MoS_2: Effect of substrate. *2D Materials, 2*, 015005.
164. Kumano, N., Inagaki, K., Kolosov, O. V., & Wright, O. B. (1998). In *IEEE Ultrasonics Symposium* (37th ed., pp. 1269–1272). IEEE.
165. Cuberes, M. T., Assender, H. E., Briggs, G. A. D., & Kolosov, O. V. (2000). Heterodyne force microscopy of PMMA/rubber nanocomposites: Nanomapping of viscoelastic response at ultrasonic frequencies. *Journal of Physics D: Applied Physics, 33*, 2347–2355.
166. Nonnenmacher, M., O'Boyle, M. P., & Wickramasinghe, H. K. (1991). Kelvin probe force microscopy. *Applied Physics Letters, 58*, 2921–2923.
167. Wu, H., et al. (2018). All-inorganic perovskite quantum dot-monolayer MoS_2 mixed-dimensional van der Waals heterostructure for ultrasensitive photodetector. *Advancement of Science, 5*, 1801219.
168. Oliver, R. A. (2008). Advances in AFM for the electrical characterization of semiconductors. *Reports on Progress in Physics, 71*, 076501.
169. Bussmann, B. K., Ochedowski, O., & Schleberger, M. (2011). Doping of graphene exfoliated on $SrTiO_3$. *Nanotechnology, 22*, 265703.
170. Melitz, W., Shen, J., Kummel, A. C., & Lee, S. (2011). Kelvin probe force microscopy and its application. *Surface Science Reports, 66*, 1–27.
171. Jacobs, H. O., Knapp, H. F., & Stemmer, A. (1999). Practical aspects of Kelvin probe force microscopy. *Review of Scientific Instruments, 70*, 1756–1760.
172. Collins, L., Kilpatrick, J. I., Kalinin, S. V., & Rodriguez, B. J. (2018). Towards nanoscale electrical measurements in liquid by advanced KPFM techniques: A review. *Reports on Progress in Physics, 81*, 086101.
173. Neubeck, S., et al. (2012). Scanning gate microscopy on a graphene quantum point contact. *Physics E, 44*, 1002–1004.
174. Mackenzie, D. M. A., Panchal, V., Corte-León, H., Petersen, D. H., & Kazakova, O. (2019). Qualitative analysis of scanning gate microscopy on epitaxial graphene. *2D Materials, 6*, 025023.
175. Sellier, H., et al. (2011). On the imaging of electron transport in semiconductor quantum structures by scanning-gate microscopy: Successes and limitations. *Semiconductor Science and Technology, 26*, 064008.
176. Brun, B., et al. (2014). Wigner and Kondo physics in quantum point contacts revealed by scanning gate microscopy. *Nature Communications, 5*, 4290.
177. Kim, K., Jeong, W. H., Lee, W. C., & Reddy, P. (2012). Ultra-high vacuum scanning thermal microscopy for nanometer resolution quantitative thermometry. *ACS Nano, 6*, 4248–4257.
178. Zhang, Y., et al. (2019). A review on principles and applications of scanning thermal microscopy (SThM). *Advanced Functional Materials, 30*, 1900892.
179. Majumdar, A. (1999). Scanning thermal microscopy. *Annual Review of Materials Science, 29*, 505–585.
180. Jóźwiak, G., Wielgoszewski, G., Gotszalk, T., & Kępiński, L. (2013). Thermal mapping of a scanning thermal microscopy tip. *Ultramicroscopy, 133*, 80–87.
181. Lienhard V, J. H., & Lienhard IV, J. H. (2018). *A heat transfer textbook* (4th ed.). Phlogiston Press.
182. Prasher, R. (2005). Predicting the thermal resistance of nanosized constrictions. *Nano Letters, 5*, 2155–2159.
183. Rohsenow, W. M., Hartnett, J. R., & Cho, Y. I. (1998). *Handbook of heat transfer* (3rd ed.). McGraw-Hill Education.
184. Goldsmid, H. J. (2010). *Introduction to thermoelectricity* (1st ed.). Springer-Verlag.
185. Schwab, K., Henriksen, E. A., Worlock, J. M., & Roukes, M. L. (2000). Measurement of the quantum of thermal conductance. *Nature, 404*, 974–977.

186. Timofeeva, M., et al. (2016). Scanning thermal microscopy with heat conductive nanowire probes. *Ultramicroscopy, 162*, 42–51.
187. Puyoo, E., Grauby, S., Rampnoux, J. M., Rouviere, E., & Dilhaire, S. (2010). Thermal exchange radius measurement: Application to nanowire thermal imaging. *Review of Scientific Instruments, 81*, 073701.
188. Harzheim, A., Evangeli, C., Kolosov, O. V., Gehring, P. (2020). Direct mapping of local Seebeck coefficient in 2D material nanostructures via scanning thermal gate microscopy. *2D Materials, 7*, 041004.
189. Cai, Y., et al. (2023). Anisotropic phonon thermal transport in two-dimensional layered materials. *Frontiers of Physics, 18*, 43303.
190. Yang, J., Liu, C., Xie, H., & Yu, W. (2021). Anisotropic heat transfer properties of two-dimensional materials. *Nanotechnology, 32*, 162001.
191. Kalantari, M. H., & Zhang, X. (2023). Thermal transport in 2D materials. *Nanomaterials, 13*, 1–32.
192. Jiang, J.-W., Zhuang, X., & Rabczuk, T. (2013). Orientation dependent thermal conductance in single-layer MoS_2. *Science and Reports, 3*, 2209.
193. Pham, V.-T., & Fang, T.-H. (2020). Effects of temperature and intrinsic structural defects on mechanical properties and thermal conductivities of InSe monolayers. *Science and Reports, 10*, 15082.
194. Tabarraei, A. (2015). Thermal conductivity of monolayer hexagonal boron nitride nanoribbons. *Computational Materials Science, 108*, 66–71.
195. Kim, S. E., et al. (2021). Extremely anisotropic van der Waals thermal conductors. *Nature, 597*, 660–665.
196. Brennan, C. J., Nguyen, J., Yu, E. T., & Lu, N. S. (2015). Interface adhesion between 2D materials and elastomers measured by buckle delaminations. *Advanced Materials Interfaces, 2*, 1500176.
197. Ding, L.-P., Shao, P., & Ding, F. (2021). Mechanism of 2D materials' seamless coalescence on a liquid substrate. *ACS Nano, 15*, 19387–19393.
198. Niu, C., Lin, F., Wang, Z. M., Bao, J., & Hu, J. (2018). Graphene levitation and orientation control using a magnetic field. *Journal of Applied Physics, 123*, 044302.
199. Kim, H. G., et al. (2017). Effect of graphene-substrate conformity on the in-plane thermal conductivity of supported graphene. *Carbon, 125*, 39–48.
200. Wang, Z. Q., et al. (2011). Thermal transport in suspended and supported few-layer graphene. *Nano Letters, 11*, 113–118.
201. Carvalho, T. C. V., et al. (2019). Temperature-dependent phonon dynamics of supported and suspended monolayer tungsten diselenide. *AIP Advances, 9*, 085316.
202. Chen, X.-K., Zeng, Y.-J., & Chen, K.-Q. (2020). Thermal transport in two-dimensional heterostructures. *Frontiers in Materials, 7*, 1–30.
203. Alborzi, M. S., Rajabpour, A., & Montazeri, A. (2020). Heat transport in 2D van der Waals heterostructures: An analytical modeling approach. *International Journal of Thermal Sciences, 150*, 106237.
204. Zhang, Z., Hu, S., Chen, J., & Li, B. (2017). Hexagonal boron nitride: A promising substrate for graphene with high heat dissipation. *Nanotechnology, 28*, 225704.
205. Wang, X., Huang, T., & Lu, S. (2013). High performance of the thermal transport in graphene supported on hexagonal boron nitride. *Applied Physics Express, 6*, 075202.
206. Chen, C. C., Li, Z., Shi, L., & Cronin, S. B. (2014). Thermal interface conductance across a graphene/hexagonal boron nitride heterojunction. *Applied Physics Letters, 104*, 081908.
207. Ma, J.-J., et al. (2019). First-principles calculations of thermal transport properties in MoS_2/$MoSe_2$ bilayer heterostructure. *Physical Chemistry Chemical Physics: PCCP, 21*, 10442–10448.
208. Gao, Y., Liu, Q., & Xu, B. (2016). Lattice mismatch dominant yet mechanically tunable thermal conductivity in bilayer heterostructures. *ACS Nano, 10*, 5431–5439.
209. Chen, Z., et al. (2023). Growth strategy for solution-phase growth of two-dimensional nanomaterials via a unified model. *Nature Synthesis, 2*, 670–677.

210. Srivastava, S., & Kotov, N. A. (2009). Nanoparticle assembly for 1D and 2D ordered structures. *Soft Matter, 5*, 1146–1156.
211. Wang, W., et al. (2022). In situ synthesis of 2D/2D MXene-COF heterostructure anchored with Ag nanoparticles for enhancing Schottky photocatalytic antibacterial efficiency under visible light. *Journal of Colloid and Interface Science, 608*, 735–748.
212. Kim, J., Byun, S., Smith, A. J., Yu, J., & Huang, J. (2013). Enhanced electrocatalytic properties of transition-metal dichalcogenides sheets by spontaneous gold nanoparticle decoration. *Journal of Physical Chemistry Letters, 4*, 1227–1232.
213. Gnatyuk, O. P., et al. (2018). 2D-BN nanoparticles as a spectroscopic marker and drug delivery system with protection properties. *RSC Advances, 8*, 30404–30411.
214. Ibrahim, K. H., et al. (2018). Simultaneous fabrication and functionalization of nanoparticles of 2D materials with hybrid optical properties. *Advanced Optical Materials, 6*, 1701365.
215. Wang, X., Sun, G., Li, N., & Chen, P. (2016). Quantum dots derived from two-dimensional materials and their applications for catalysis and energy. *Chemical Society Reviews, 45*, 2239–2262.
216. James Singh, K., et al. (2021). Recent advances in two-dimensional quantum dots and their applications. *Nanomaterials, 11*, 1–30.
217. Shinde, P. V., Tripathi, A., Thapa, R., & Sekhar Rout, C. (2022). Nanoribbons of 2D materials: A review on emerging trends, recent developments and future perspectives. *Coordination Chemistry Reviews, 453*, 214335.
218. Sinha, S. S., Sreedhara, M. B., & Tenne, R. (2021). Why do nanocrystals of 2D materials form nanotubes and why is that important? *Nano Today, 37*, 101060.
219. Berber, S., Kwon, Y. K., & Tomanek, D. (2000). Unusually high thermal conductivity of carbon nanotubes. *Physical Review Letters, 84*, 4613–4616.
220. Aftab, S., Iqbal, M. Z., & Rim, Y. S. (2023). Recent advances in rolling 2D TMDs nanosheets into 1D TMDs nanotubes/nanoscrolls. *Small (Weinheim an der Bergstrasse, Germany), 19*, 2205418.
221. Kim, J. H., Pham, T. V., Hwang, J. H., Kim, C. S., & Kim, M. J. (2018). Boron nitride nanotubes: Synthesis and applications. *Nano Convergence, 5*, 17.
222. Tans, S. J., et al. (1997). Individual single-wall carbon nanotubes as quantum wires. *Nature, 386*, 474–477.
223. Cao, Y., et al. (2018). Unconventional superconductivity in magic-angle graphene superlattices. *Nature, 556*, 43–50.
224. Cao, Y., et al. (2018). Correlated insulator behaviour at half-filling in magic-angle graphene superlattices. *Nature, 556*, 80–84.
225. Hennighausen, Z., & Kar, S. (2021). Twistronics: A turning point in 2D quantum materials. *Electron Structure., 3*, 014004.
226. Yankowitz, M., et al. (2019). Tuning superconductivity in twisted bilayer graphene. *Science, 363*, 1059–1064.
227. Xu, Q., Guo, Y., & Xian, L. (2022). Moiré flat bands in twisted 2D hexagonal vdW materials. *2D Materials, 9*, 014005.
228. Alexeev, E. M., et al. (2019). Resonantly hybridized excitons in Moiré superlattices in van der Waals heterostructures. *Nature, 567*, 81–86.
229. Yoo, H., et al. (2019). Atomic and electronic reconstruction at the van der Waals interface in twisted bilayer graphene. *Nature Materials, 18*, 448–453.
230. Li, Y., et al. (2021). Unraveling strain gradient induced electromechanical coupling in twisted double bilayer graphene Moiré superlattices. *Advanced Materials, 33*, 2105879.
231. Ryu, Y. K., Frisenda, R., & Castellanos-Gomez, A. (2019). Superlattices based on van der Waals 2D materials. *Chemical Communications, 55*, 11498–11510.
232. Zhao, P., Xiao, C., & Yao, W. (2021). Universal superlattice potential for 2D materials from twisted interface inside h-BN substrate. *npj 2D Materials and Application, 5*, 38.
233. Twaha, S., Zhu, J., Yan, Y., & Li, B. (2016). A comprehensive review of thermoelectric technology: Materials, applications, modelling and performance improvement. *Renewable Sustainable Energy Review, 65*, 698–726.

234. Das, A., & Banerji, P. (2023). Unusual transport and impact of nonparabolic electronic band structure on the thermoelectric performance in n-type In_4Se_3-based thermoelectric materials. *Physica Status Solidi B, 260*, 2300078.
235. Xu, Y., Li, Z. Y., & Duan, W. H. (2014). Thermal and thermoelectric properties of graphene. *Small (Weinheim an der Bergstrasse, Germany), 10*, 2182–2199.
236. Ibañez-Puy, M., Bermejo-Busto, J., Martín-Gómez, C., Vidaurre-Arbizu, M., & Sacristán-Fernández, J. A. (2017). Thermoelectric cooling heating unit performance under real conditions. *Applied Energy, 200*, 303–314.
237. Graus, W. H. J., Voogt, M., & Worrell, E. (2007). International comparison of energy efficiency of fossil power generation. *Energy Policy, 35*, 3936–3951.
238. Wibisono, A. F., & Shwageraus, E. (2016). Thermodynamic performance of pressurized water reactor power conversion cycle combined with fossil-fuel superheater. *Energy, 117*, 190–197.
239. Roy, D., Samanta, S., & Ghosh, S. (2020). Performance assessment of a biomass fuelled advanced hybrid power generation system. *Renewable Energy, 162*, 639–661.
240. Letcher, T. M. (2016). *Storing energy with special reference to renewable energy sources* (1st ed.). Elsevier.
241. Schygulla, P., et al. (2022). In *International conference on metal organic vapor phase epitaxy* (20th ed.). n/a, Fraunhofer-Gesellschaft.
242. Zarrouk, S. J., & Moon, H. (2014). Efficiency of geothermal power plants: A worldwide review. *Geothermics, 51*, 142–153.
243. Jiang, B., et al. (2022). High figure-of-merit and power generation in high-entropy GeTe-based thermoelectrics. *Science, 377*, 208–213.
244. Chiritescu, C., et al. (2007). Ultralow thermal conductivity in disordered, layered WSe_2 crystals. *Science, 315*, 351–353.
245. Gayner, C., & Amouyal, Y. (2020). Energy filtering of charge carriers: Current trends, challenges, and prospects for thermoelectric materials. *Advanced Functional Materials, 30*, 1901789.
246. Dresselhaus, M. S., et al. (2007). New directions for low-dimensional thermoelectric materials. *Advanced Materials, 19*, 1043–1053.
247. Wu, W., et al. (2014). Piezoelectricity of single-atomic-layer MoS_2 for energy conversion and piezotronics. *Nature, 514*, 470–474.
248. Aized, T., Khan, M. B., Raza, H., & Ilyas, M. (2017). Production routes, electromechanical properties and potential application of layered nanomaterials and 2D nanopolymeric composites—A review. *International Journal of Advanced Manufacturing Technology, 93*, 3449–3459.
249. Lowell, J., & Rose-Innes, A. C. (1980). Contact electrification. *Advances in Physics, 29*, 947–1023.
250. Yang, P. K., & Lee, C. P. (2019). 2D-layered nanomaterials for energy harvesting and sensing applications. In A. El-Shahat & M. Ruba (Eds.), *Applied electromechanical devices and machines for electric mobility solutions* (Chap. 11, pp. 1–14). IntechOpen.
251. Denning, D., Guyonnet, J., & Rodriguez, B. J. (2016). Applications of piezoresponse force microscopy in materials research: From inorganic ferroelectrics to biopiezoelectrics and beyond. *International Materials Reviews, 61*, 46–70.

Chapter 3
Applied Methods for Materials and Structures Preparation

Abstract This chapter details the experimental procedures employed to fabricate the two-dimensional (2D) materials and device architectures investigated throughout the thesis. The focus is on accessible and reproducible methods that allow for the creation of nanostructures tailored to specific research objectives. The chapter begins with an overview of the mechanical exfoliation techniques used to isolate individual layers of various 2D materials, including graphene, hexagonal boron nitride, indium selenide and molybdenum disulfide. Each exfoliation process is preceded by precise substrate preparation to ensure optimal adhesion and flake quality. Following exfoliation, the fabrication of heterostructures is described, highlighting the importance of alignment, stacking order and interface compatibility. The chapter then covers the assembly of nanodevices, including etching strategies for accessing internal layers and methods for depositing metallic contacts with minimal contamination. To further refine device cross-sections and enable advanced characterisation, beam exit cross-sectional polishing is introduced as a complementary preparation technique. Altogether, this chapter outlines a complete set of fabrication protocols designed to preserve the intrinsic properties of 2D materials while enabling their integration into functional systems. These procedures form the experimental foundation for the investigations of thermoelectric, thermal and electromechanical phenomena presented in the subsequent chapters.

3.1 Introduction

The first step in the majority of research processes is preparing the sample that has to be analysed, especially, vdW materials require multiple predevelopment stages to convert them from bulk crystals to nanoscale thin layers. As explained in the previous sections, they require a substrate suitable for their proper support and attachment. In this chapter, the principal fabrication methodologies employed in this thesis are discussed and described. To start, the different mechanical exfoliation procedures implemented on each of the vdW materials are elucidated, given that each structure has different properties affecting the exfoliation performance. Afterwards, the steps

S. Gonzalez-Munoz, *Interplay of Nanoscale Physical Properties in 2D Materials and Heterostructures*, Springer Theses, https://doi.org/10.1007/978-3-032-14271-9_3

utilised to fabricate stacked vdW heterostructures are extensively described. Lastly, once the layers are deposited on a substrate, further advanced preparation techniques are discussed, presenting approaches for device fabrication and access to the interior layers of the samples.

All the fabrication described in this chapter has been conducted by myself in the Lancaster University Physics Department facilities unless otherwise specified. Nevertheless, the initial training and help provided by some colleagues must be acknowledged. Firstly, Prof. Oleg V. Kolosov provided me with training and continued aid in the majority of the subsequent techniques, especially the heterostructure preparation. Dr. Eli G. Castanon and Mohammadali Razeghi provided functional training with the mechanical exfoliation processes, Dr. Kunal Lulla and Dr. Yuezhen Lu mentored myself in all the cleanroom procedures and Andy Niblett and Dr. Yue Chen transferred their knowledge and gave continuous support with the beam exit cross-sectional polishing (BEXP) methodology.

3.2 Mechanical Exfoliation

3.2.1 Substrate Preparation

Given the atomically thin nature of 2D materials, a proper substrate must be selected for a number of reasons. First, the roughness of the substrate affects the flatness of the deposited material, given that, on rougher substrates, the possibility of deposition of interstitial external molecules increases, as well as the presence of air or humidity in the interlayer voids, ultimately decreasing the conformity between the two adjacent surfaces. Second, the optical compatibility between the layered material and the supporting structure is a matter of importance due to the quasi-transparent behaviour of such vdW layers to visible light, making it extremely challenging to distinguish them if they are not deposited on the proper surface [1]. Lastly, the composition of the substrate itself can also affect the properties of the 2D specimen and its interface.

For these reasons, 90 or 300 nm thermally grown SiO_2 on silicon layers are generally the substrates of choice, given its chemically inert and isolating nature to both thermal and electrical transport (in the case of the SiO_2 layer), their adaptability and cheap manufacturing cost, often employing such combination in the complementary metal-oxide semiconductor (CMOS) industry and their enhanced interference optical effect of the SiO_2/silicon stacked structure, improving the 2D flake visualisation [2, 3]. Other materials can also be suitable substrates depending on the intended functionality of the sample, with silicon, silicon nitrides, aluminium oxides, mica, copper, gold or flat polymers being the most utilised (Fig. 3.1) [4].

The majority of substrates can be purchased as standardised wafers of 2, 3, 4, 6, 8, 12 or 18-inch sizes, which are often too large to fit in the measurement stages. Hence, smaller pieces of these wafers, called chips, must be prepared as the final substrate with suitable lateral dimensions. Specifically, the chips employed in this project have

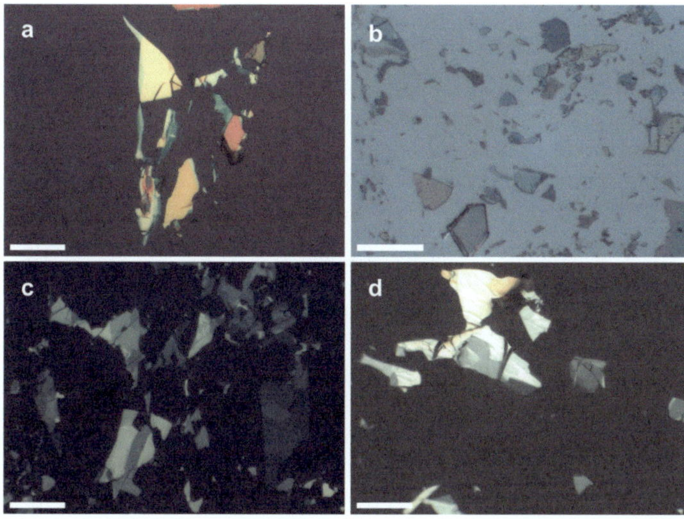

Fig. 3.1 Different contrast of γ-InSe flakes on **a** SiO_2, **b** silicon, **c** silicon nitride (Si_3N_4) and **d** PDMS substrates. Notice that the flakes on SiO_2 show a clearer contrast difference, being able to identify their qualitative thicknesses by the flake's colour. Scale bars are 30 μm

areas of 0.50×0.50, 1×1 or 2×2 cm^2 depending on the further fabrication and characterisation steps and are cut from standard 4-inch wafers (Fig. 3.2a).

Given that the wafer materials are crystalline, i.e., with an ordered structure, the precision and easiness of the cuts increase if they are performed along the crystal

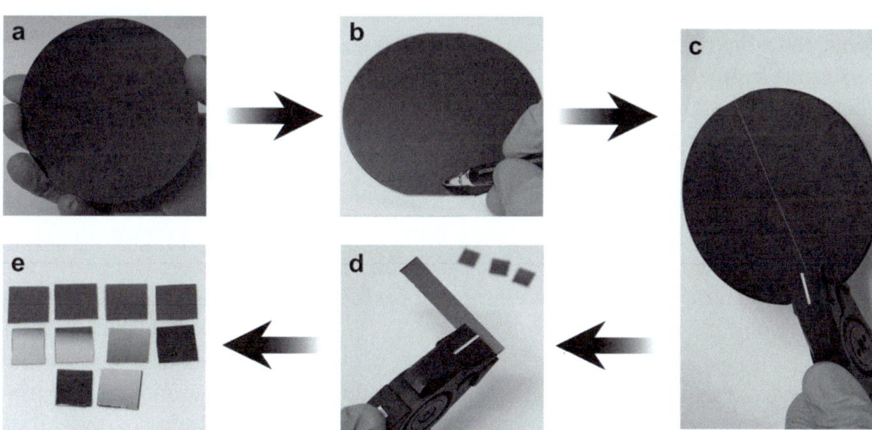

Fig. 3.2 Wafer-cleaving steps. **a** Pristine SiO_2 4-inch wafer, **b** diamond-tip pen scratches a line on the dull face of the wafer, following the crystallographic direction, **c** wafer-cleaving pliers cut the wafer along the marked line, **d** the same procedure is repeated to break the wafer into **e** the desired dimensions

preferential directions. Additionally, one of the faces of said wafers is pre-polished by the manufacturer to increase its quality and to indicate the side which has to be used at the subsequent fabrication phases as a surface with minimised roughness. Hence, to avoid contamination and achieve a straight cleavage, a line is scratched with a diamond-tip pen on the dull face along the crystallographic direction referenced by the primary and secondary perimeter flats (Fig. 3.2b). Afterwards, a set of wafer-cleaving pliers are positioned over the drawn line, and its jaws are gently pressed until the crystal breaks following the line direction (Fig. 3.2c). This process is then iterated marking the desired dimensions aided with a ruler and cleaving with the same pliers (Fig. 3.2d). Finally, the chips are reduced to the specific dimensions (Fig. 3.2e) and are ready to be cleaned.

In order to have a proper adhesion of the vdW materials on the substrate, it is recommended to treat the supporting material prior to any further deposition steps. This pretreatment decreases the amount of contamination and humidity on the surface, which is a source of defects and roughness, and activates its chemical bindings, resulting in stronger bonds. The range of pollutants adsorbed on these samples can include both inorganic and organic chemical compounds, however hydrocarbons, carboxyl groups, water as well as nitrogen and oxygen terminations are the most common passivation layers that hinder the proper attachment between the 2D material and its supporting structure. These may originate from prior processes such as wafer manufacturing, transport, or storage. Additionally, the cleaving process releases dust particles of the same substrate that can also be deposited on the surface. For this reason, any treatment steps involving the increase of functionalisation of the substrates should be performed in controlled environments with reduced pollution agents, such as a cleanroom or, ideally, a glovebox containing an inert gas like argon.

Figure 3.3 depicts the standard preparation steps followed here. Firstly, an ultrasonication bath of the chip in acetone is conducted for 10 min at RT and maximum power, repeating the process in IPA and DIW (Fig. 3.3a). The chip must be rapidly transferred from acetone to IPA since the former tends to dry fast, leaving organic residues. If the substrate is highly contaminated, these baths can be conducted at higher temperatures (50–60 °C) and rinsing the sample after each sonication with the next solvent to be used. Moreover, an initial sonication in EtOH can be beneficial. After the sonication baths, the sample is thoroughly dried with an inert gas such as nitrogen or argon (Fig. 3.3b), removing most of the residues trapped in the DIW solvent. Lastly, the face of the substrate on which the deposition has to be performed is exposed to an oxygen plasma treatment (Fig. 3.3c) via two different reactive ion etchers for surface activation (see Sect. A.2.2). The highly energetic ionic plasma impinges with the surface, chemically reacting with any remaining organic particles as well as disrupting their bindings, and is evacuated from the chamber by a continuous gas flow carrying away the residues and leaving pristine and activated substrate terminations. Afterwards, the substrate is extracted from the chamber to proceed with the deposition steps (Fig. 3.3d), considering the discussion in Sect. A.2.1.

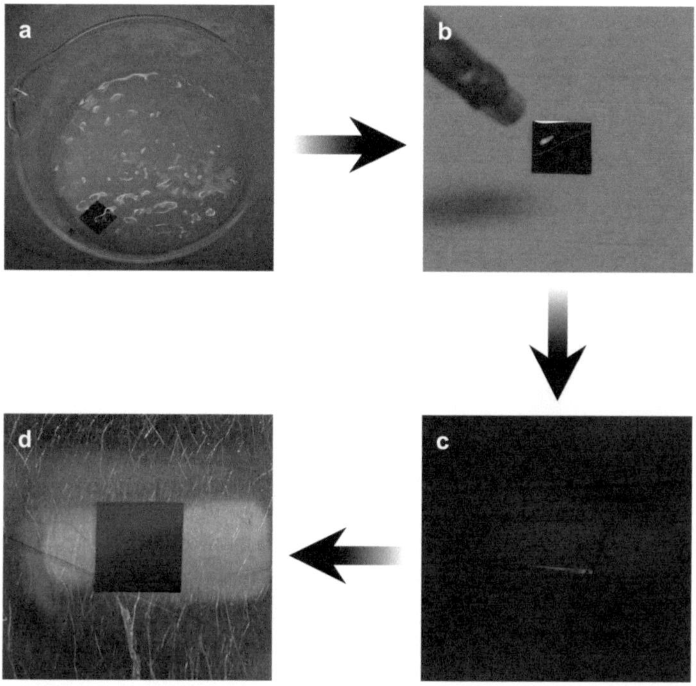

Fig. 3.3 Substrate cleaning steps. **a** A SiO_2 substrate chip is ultrasonicated inside a beaker containing acetone, the process is then repeated with IPA and DIW, **b** the chip is dried with a nitrogen gas flow, **c** the chip is exposed to an oxygen plasma etching and **d** the final cleaned and activated chip is extracted and ready for further deposition

3.2.2 Graphene and h-BN Exfoliation

To separate thin graphene layers from their bulk graphite counterpart, the following steps are used. These steps can be applied to h-BN and all the different types of graphite due to their similar mechanical properties, varying the exfoliation process only in the size of the bulk crystals and their macroscopic contrast identification (Fig. A.1a–d) [5]. Also, the bulk crystals should be kept under vacuum when they are not used to avoid oxidation. A common practice is to exfoliate and discard both topmost layers from the top and bottom faces of the 3D structure, exposing fresh and pristine areas. This consideration can be applied to all the materials used in this work.

First, a 10×1 cm^2 piece of Nitto ELP BT-130E-SL blue tape is cut, and the bulk graphite crystal is placed on its centre, then, by folding and pressing the tape against the faces of the crystal for ~ 2 s and peeling it at a medium pace (Fig. 3.4a), multiple thick flakes remain on the tape after the bulk crystal is removed (Fig. 3.4b). The folding-peeling step is repeated 5–6 times with the remaining flakes on clean areas of the tape until the flakes are almost transparent, contacting only the freshest exfoliated areas. Hence, the flakes further from the tape's centre are the thinnest

(Fig. 3.4c and d). A smaller (3×1 cm^2) piece of blue tape is placed covering the last exfoliated area and peeled from it, transferring the thinnest flakes (Fig. 3.4e). After the substrate is prepared as explained in Sect. 3.2.1, the tape and flakes are deposited on the activated surface. The attachment is improved by softly rubbing the backside of the tape with a cotton swab, dragging away any big trapped air bubbles (Fig. 3.4f) and annealing on a hot plate for 10 min at 80 °C, removing the majority of the stuck humidity (Fig. 3.4g). Lastly, the tape is removed at a slow pace and a small angle from the surface (Fig. 3.4h). In this way, multiple flakes of different thicknesses and sizes (identified qualitatively under the optical microscope in Fig. 3.5) remain on the substrate without transferring adhesive contaminants.

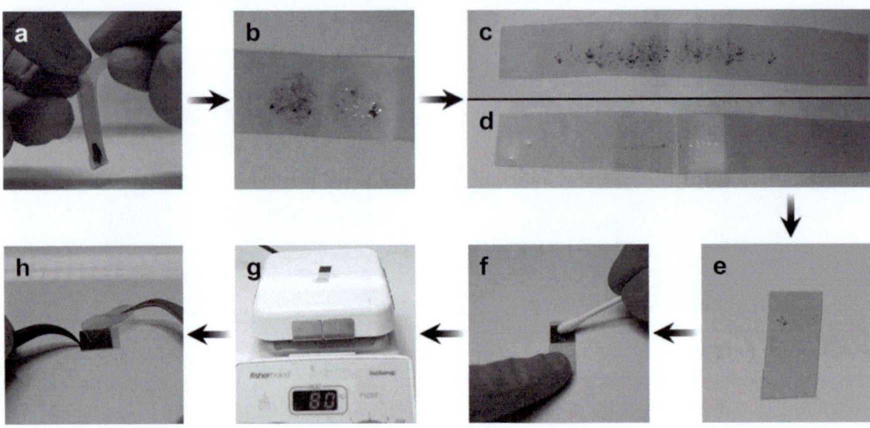

Fig. 3.4 Graphene and h-BN exfoliation steps. **a** Bulk graphite crystal pressed and peeled from the blue tape, **b** graphite flakes after removing the bulk crystal from the tape, remaining **c** graphite and **d** h-BN flakes after multiple tape folds, **e** last piece of tape with thin graphite flakes, **f** tape and flakes being attached to the substrate, **g** ensemble on the hot plate for improved adhesion and **h** tape peeling from the substrate

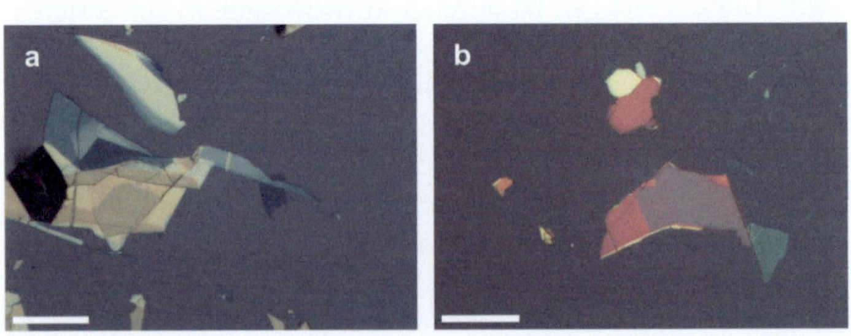

Fig. 3.5 Optical images of exfoliated **a** graphene and **b** h-BN. Notice the different colours of the flakes indicating their thickness. Specifically, a yellow contrast represents high thickness, pink represents middle thickness and blue represents low thickness, with darker blue being the thinnest and brighter yellow the thickest [5]. Scale bars are 70 μm

3.2.3 InSe Exfoliation

As a post-TMD with a small interlayer binding energy, InSe is easier to separate from the 3D crystal than regular TMDs, hence, an exfoliation process similar to that of graphite can be applied. However, some steps differ since InSe has a higher tear tendency compared to graphene or h-BN, forcing to reduce the number of peeling stages. The same separation procedures can be applied to both γ-InSe and α-In_2Se_3 due to their comparable mechanical stability.

After peeling both faces of the bulk crystal on a preliminary Nitto ELP BT-150E-KL mother tape (Fig. 3.6a), a smaller piece of the same type is pressed for 2 s (Fig. 3.6b) and peeled (Fig. 3.6c) to remove thinner flakes from the initial ribbon and transfer them to a new fixed 10×1 cm^2 piece of blue strip (Fig. 3.6d). Then, the pressing and peeling process is repeated on clean areas of the fixed tape between 4 and 5 times, always peeling at a medium pace and large angle from the movement direction. Each exfoliation thins down the flakes until they become quasi-transparent (Fig. 3.6e). The area with the last exfoliation performed on the fixed tape is attached (Fig. 3.6f) and pressed for 30 s (Fig. 3.6g) on the activated surface of a fixed substrate, to be subsequently peeled off slowly (Fig. 3.6h) as in the graphene exfoliation, transferring the flakes to the substrate (Fig. 3.7).

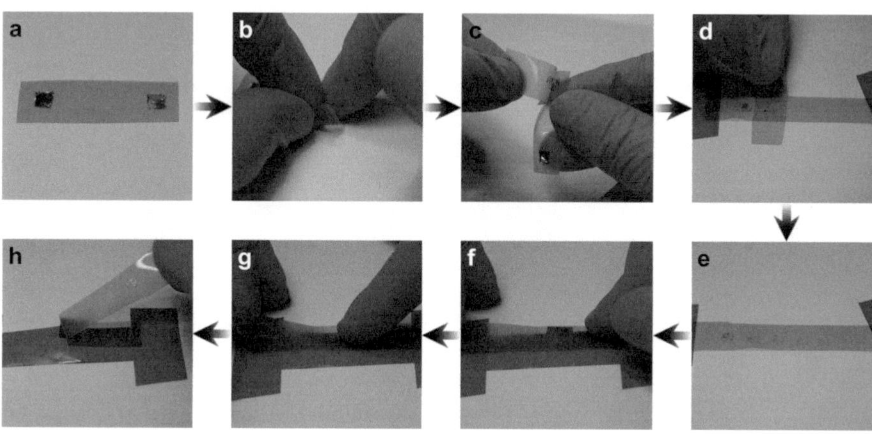

Fig. 3.6 InSe exfoliation steps. **a** Bulk γ-InSe crystal preliminary exfoliation on the first blue tape, second piece of blue tape **b** pressed and **c** peeled from the thin preliminary exfoliation, **d** transfer of the exfoliated flakes from the second tape to a fixed third blue tape, **e** flakes' multiple thinning steps on the fixed tape, **f** thin flakes on the tape attached to a fixed substrate, **g** tape pressed on the substrate and **h** tape peeling

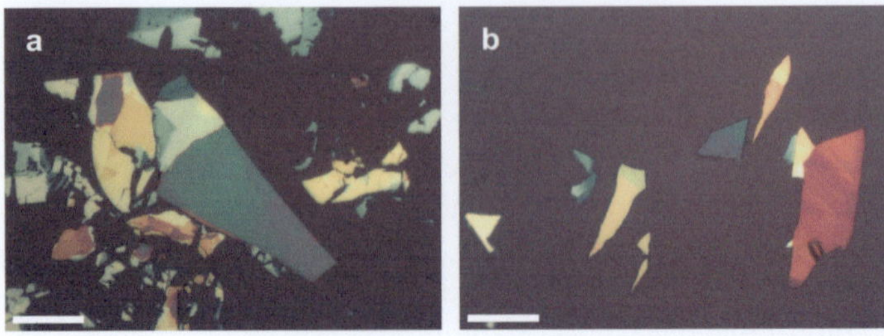

Fig. 3.7 Optical images of exfoliated **a** γ-InSe and **b** α-In_2Se_3. Scale bars are 30 μm

3.2.4 MoS_2 Exfoliation

TMDs are one of the most challenging materials to mechanically exfoliate, given their high binding energy combined with a low shear modulus and an elevated edge defect quantity [6], causing a limitation in the number of peeling steps that can be performed without notoriously decreasing the flakes' lateral dimensions (Fig. 3.9). This means that the probabilities of exfoliating a large-surface monolayer employing the regular method are low. Hence, the next method has been adapted to yield the best quality possible of thin and moderately large MoS_2 flakes on SiO_2 substrates. Nevertheless, these steps can be applied to the majority of TMDs (Fig. 3.8).

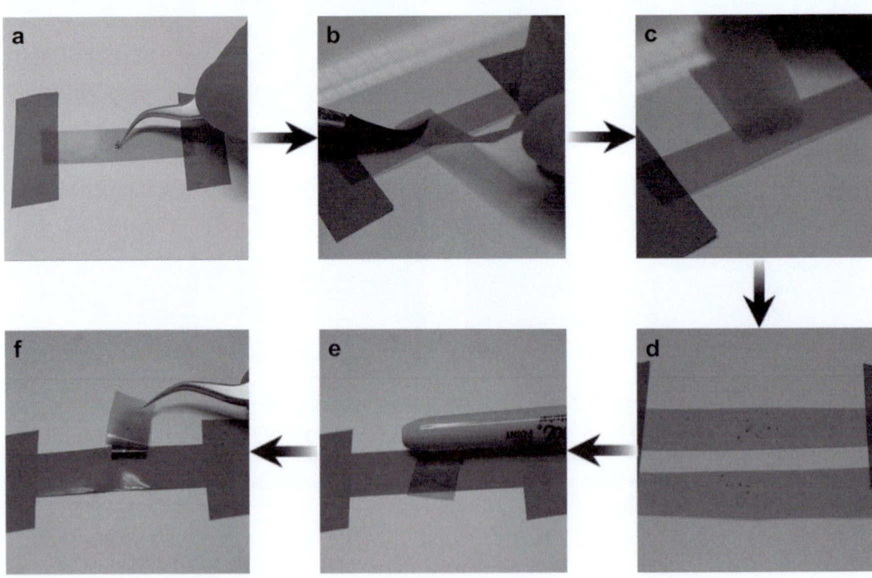

Fig. 3.8 MoS_2 exfoliation steps. **a** Small piece of bulk MoS_2 crystal tapped on a fixed blue tape, second piece of folded blue tape **b** attached and **c** peeled from the first tape, **d** remaining flakes on the first and second tape, **e** cut piece of the second tape attached and flattened on a fixed substrate and **f** tape peeled from the fixed substrate

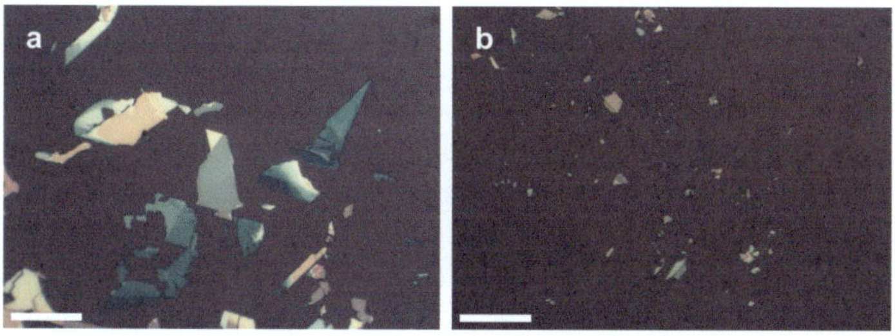

Fig. 3.9 Optical images of MoS_2 flakes after **a** three and **b** six exfoliation folds. Notice the reduction of lateral size after doubling the folding steps. Scale bars are 30 μm

To start, a 3 × 1.50 cm^2 strip of Nitto ELP BT-150E-KL is fixed on a flat surface. Then, a small piece of MoS_2 extracted from the bulk crystal is tapped gently (Fig. 3.8a) several times on the tape until an area of approximately half of the size of the substrate is covered with thick flakes. Afterwards, a fresh blue tape of the same dimensions is attached to its centre contacting with the mother flakes by a gentle rubbing with the tweezers (Fig. 3.8b) and peeled at a medium pace (Fig. 3.8c). This step is repeated on fresh areas of both tapes for a maximum of three times, filling an area of approximately the size of the substrate with MoS_2 flakes (Fig. 3.8d). With the activated substrate fixed, one can carefully deposit the exfoliated flakes by attaching the last tape and continuously rubbing its back side with a flat surface for 2 min (Fig. 3.8e). Finally, the tape can be removed as slowly as possible and in a continuous movement from the substrate (Fig. 3.8f).

3.3 Fabrication of Heterostructures

The precise manipulation and stacking of individual 2D layers is a process that allows for a deterministic deposition of materials forming tunned homo- or heterostructures with a high yield. However, due to the low stability of 2D materials, a polymeric matrix is required as a transfer substrate, aiding in the procedure while introducing contamination which can affect the ensemble's performance. Hence, engineering of the employed polymers is paramount [7].

Three stacking techniques are tested with the target of obtaining the cleanest, thinnest and largest heterostructure ensemble. PMMA film carrier method is assessed with graphene layers, however, the requirement of strong chemical solvents and the low mechanical stability of PMMA increase the contamination and the cracks on the final structure. Two variations of microcontact stamping are also tested. The first one is the classic PDMS/PPC/h-BN hot pick-up transfer method, where the polymeric layers are synthesised and treated following the regular procedure [8, 9]. However, the PDMS/PPC interlayer strength is insufficient to hold the PPC layer on the polymer stamp after the transfer, leaving a large amount of PPC contaminant on the sample.

Moreover, the high temperatures required and the large number of steps decrease the efficiency of the process. In this regard, the second microstamping variation of conventional PDMS drop-down represents a much simpler and straightforward assembly technique [10], being the procedure adopted to create the heterostructures in this thesis. It must be mentioned that this process is not suitable for 1LG or few-layer h-BN due to the low contrast in the identification steps, nevertheless, it can be applied for transferring monolayer TMDs, FLG and InSe.

The process starts by exfoliating the 2D material on the corresponding blue tape as depicted in Sect. 3.2 and covering the resulting flakes with a new piece of blue tape (Fig. 3.10a) while applying slight pressure with a cotton swab. Then, a 1.50 $\times$ 1 cm^2 rectangular piece of Gel-Pak® Gel-Film® PF-6.5-X4 is cut and peeled from its two protective layers (Fig. 3.10b). Once the PDMS layer is freestanding (Fig. 3.10c), it can be transferred with tweezers close to the edge of a previously cleaned glass slide (Fig. 3.10d), such transfer has to be performed carefully in order to avoid leaving wrinkles or air bubbles between the slide and the polymers. The blue tape/2D material/blue tape sandwich-like ensemble can be peeled and brought into contact with the PDMS layer (Fig. 3.10e) to be sequentially peeled at a rapid sequence (Fig. 3.10f), leaving several flakes on the PDMS stamp (Fig. 3.10g).

In order to selectively transfer the flakes onto a target substrate, an accurate positioning system, known as transfer station, is needed. Such system allows to move both the sample and the polymer/flake stamp with micrometric precision. The transfer station employed herein consists of four different custom build elements (Fig. 3.11a). The first one is a fixed Mitutoyo FS70ZS optical microscope with three long working distance objectives (Mitutoyo M Plan APO 2 $\times$, 10 $\times$ and 50 $\times$). The long working distance is required to ensure enough available space between the lens and the stage and allow suitable vertical movement of the clamped glass slide. The microscope incorporates a polariser, an analyser and a differential interference contrast (DIC)

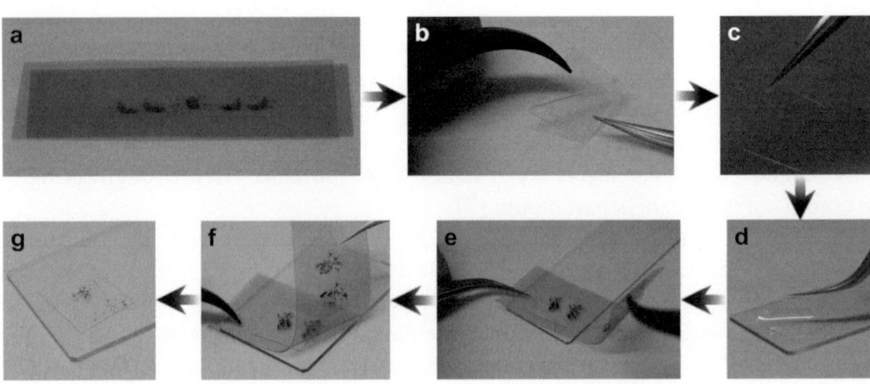

Fig. 3.10 PDMS stamp preparation. **a** 2D material exfoliated on blue tape, **b** PDMS layer peeled from the protective base, **c** freestanding PDMS layer, **d** PDMS layer deposited on a glass slide, **e** exfoliated 2D flakes deposited on the PDMS layer, **f** peeling of the blue tape and **g** transferred 2D flakes onto the PDMS stamp

unit in order to identify thin flakes with low contrast. A custom-made sample stage is fixed under the microscope. This stage is movable with a μm accuracy in the x and y directions and allows to manually change the respective angles along the x, y and z spatial axes. Also, it has three holes in its centre connected to a vacuum pump that are used to fix the substrate (Fig. 3.11b) and has an RS PRO PT1000 RTD sensor and two parallel Caddock Electronics MP915 10 Ω power film resistors attached and connected to a Meerstetter Engineering TEC-1091-PT1000-SCREW OEM precision Peltier temperature controller (TEC) which increases and monitors the temperature of the stage up to 200 °C. The remaining element is a mechanical clamp connected to a micromanipulator. This clamp holds the glass slide with the polymer stamp facing down (Fig. 3.11c) and is capable of moving it in the x, y and z directions as well as tilting the slide along the x and y axes.

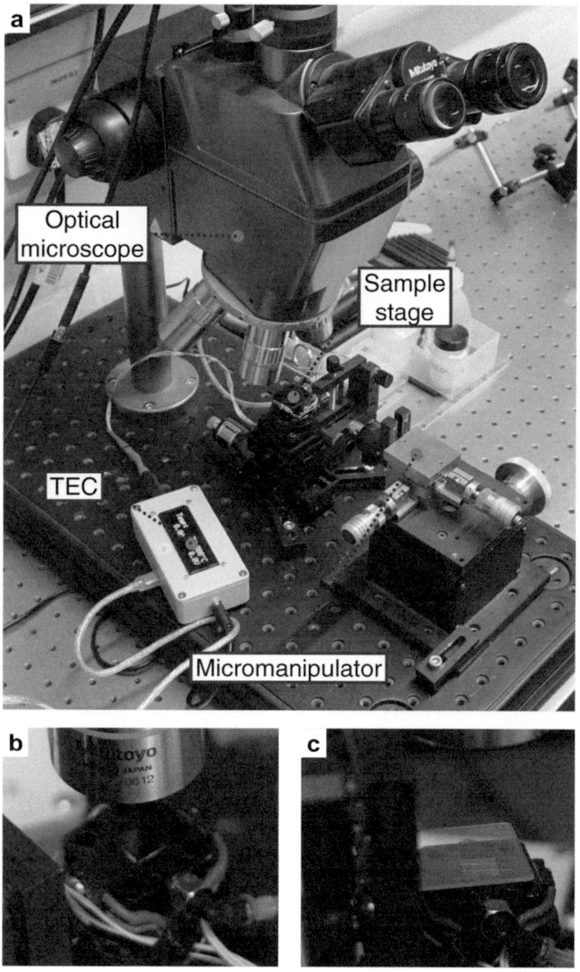

Fig. 3.11 **a** Transferring setup with its main parts highlighted (TEC stands for temperature controller), **b** close-up view of the vacuum fixed sample on the heated stage and **c** close-up view of the clamped glass slide with the polymer stamp over the sample

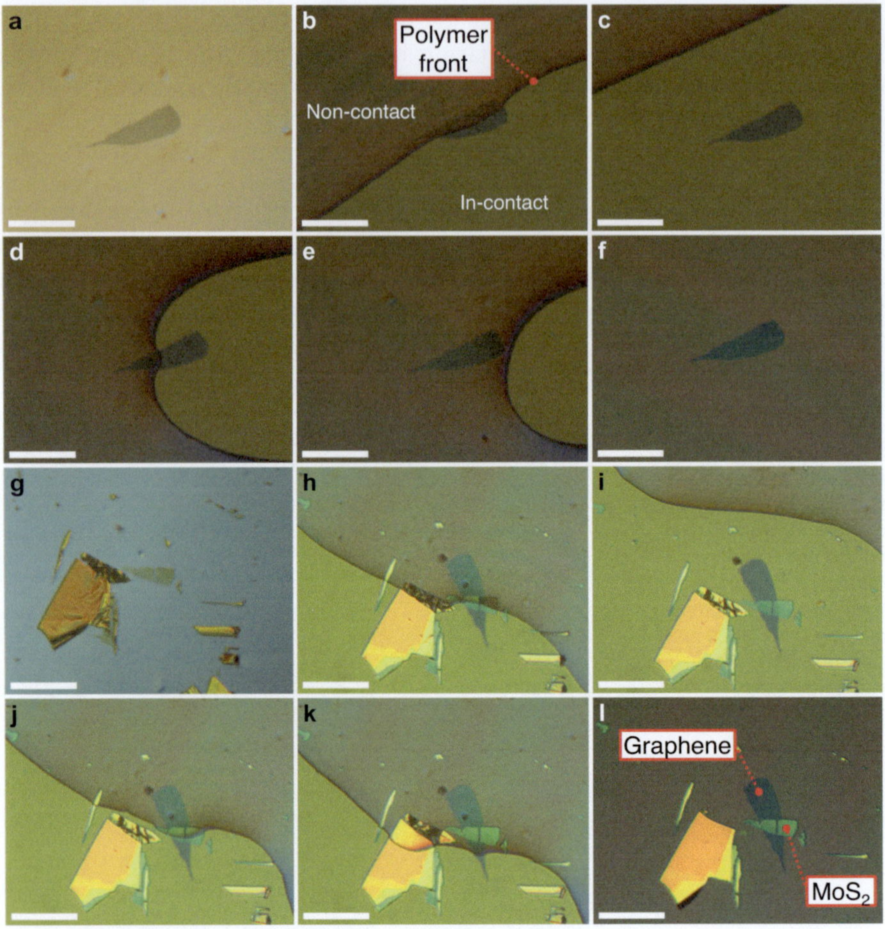

Fig. 3.12 Transferring steps for the synthesis of a graphene/MoS_2 heterostructure. Graphene deposition: **a** exfoliated graphene on PDMS, **b** graphene/PDMS first contact with a SiO_2 substrate, notice that the diagonal line is the polymer front, separating the in-contact (bright) and non-contact (dark) areas, **c** flake fully in contact with the substrate, **d** receding polymer over the flake, **e** polymer front is away from the flake and **f** graphene flake deposited on the SiO_2 substrate. MoS_2 deposition: **g** exfoliated MoS_2 on PDMS, **h** MoS_2/PDMS first contact with the graphene/SiO_2 ensemble, **i** flakes fully in contact with the ensemble, **j** receding polymer over the flakes, **k** polymer front is away from the MoS_2 flake and **l** final MoS_2/graphene heterostructure. Scale bars are 70 μm

Before starting the transferring process, the substrate is cleaned and activated as explained in Sect. 3.2.1 to increase the yield of the stamping process. The transfer process begins once the substrate and the glass slide are fixed on the station with the flakes on the polymer facing towards the activated sample surface. First, a graphene flake on the PDMS stamp is identified via the optical microscope (Fig. 3.12a) and, if needed, centred over a specific area of the substrate. Afterwards, the polymer stamp is lowered and carefully brought in contact with a SiO_2 substrate, ensuring that the whole flake is pressed to the flat surface. At this stage, the sample and the polymer may not be perfectly parallel, meaning that an edge or corner of the stamp will contact first with the substrate, this can cause the glass slide to break if it is kept lowering since no contact on the focused area will be observed if the angle between the polymer and the substrate is not extremely low. To align the two elements, the tilting angle of the stage should be changed. The first contact is shown in Fig. 3.12b, where the polymer front can be seen separating the image diagonally with in-contact (bright) and non-contact (dark) areas. When the entire flake is pressed against the surface (Fig. 3.12c), the temperature of the stage is risen to 65 °C for 2 min and cooled down to RT, enhancing the interlayer mechanical stability between the flake and the substrate. Then, the glass slide is moved up extremely slowly, causing a precise receding of the polymer front over the flake (Fig. 3.12d) and when the in-contact area surpasses the flake (Fig. 3.12e), the slide can be elevated faster to achieve the deposition of the graphene flake on the SiO_2 (Fig. 3.12f). The process is repeated with the deposition of a MoS_2 thin flake, ensuring that the TMD is centred with the deposited graphene (Fig. 3.12g–l).

Additionally, a fresh PDMS stamp can be used to precisely pick-up flakes which have been exfoliated to a solid substrate, resulting in another tool for heterostructure building. However, the receding step should be performed as fast as possible and in a single "snap" movement to ensure the flake is removed from the substrate. After the stamping procedure, the whole sample is annealed inside a vacuum oven at 160 °C for 2 h. This is done to increase the contact area and strength at the interfaces of the flakes, given that any remaining contaminants or humidity trapped in such interfaces tend to aggregate in big clusters instead of being scattered over the entire range of the interface, leaving bigger clean areas. The process can be repeated to stack different layers of materials forming thicker structures, however, the possibility of failure increases due to the presence of cracks and folds in the flakes, which are sources of tearing.

3.4 Assembly of Nanodevices

A nanodevice is defined by a particular customised geometry and a signal linkage with the macroscopic world, generally established by nanopatterning the substrate and/or the layered structure and contacting the target material via metallic electrodes, respectively, to perform a preestablished number of physical actions, such as physical sensing, actuation or transport at the nanoscale.

3.4.1 Etching

Following a top-down procedure, pattern etching can be implemented on substrates, coated layers and 2D structures to create a specific shape with defined dimensions.

This technique consists of contacting the target area of the material with a highly energetic, reactive or hard element that removes or changes the chemistry of the exposed region. The lithographic techniques explained in Sects. 2.3.3–2.3.5 can all be used as mechanical, electronic, thermal and optical etching, comprising the available types of pattering techniques together with chemical and plasma etching.

This section describes the two patterning processes utilised to etch photoresist layers and standard substrates (SiO_2, Si_3N_4 and silicon), creating holes with μm dimensions with the ultimate purpose of depositing suspended vdW layers. All the processes were conducted inside the Lancaster University Physics Department QTC Class 1000 and 100 cleanrooms.

To start, the substrate has to be treated as explained in Sect. 3.2.1 before it is spin coated. Also, it is beneficial to prebake the substrate for 1 min at 135 °C to increase the adhesion. The coating, performed with a SÜSS MicroTec LabSpin6, is composed of two photoresist layers. First, the substrate is fixed on the vacuum chuck stage and covered entirely avoiding the transfer of bubbles with liquid MicroChem LOR 3A resist dispensed with a pipette (Fig. 3.13a) to be subsequently spun at 3000 rpm for 30 s (Fig. 3.13b) and baked at 180 °C for 5 min, flattening the resist on the μm range. The second resist is added by coating a layer of MICROPOSIT™ S1813™ at 6000 rpm for 1 min and baking the ensemble at 115 °C for 2 min. The corners of the sample are flattened with a swab moistened in acetone since thicker areas of resist tend to accumulate there during the spin coating.

Afterwards, the assembly is introduced into a SÜSS MicroTec MJB4 mask aligner to expose the photoresists to UV light (wavelength of 365 nm) through a prepatterned chromium mask, imprinting the pattern on the resists by breaking their intralayer bonds (Fig. 3.13c). The resolution of the optical lithography can be enhanced by modifying the contact and exposure times (20 and 2.40 s, respectively) and by pressing the substrate to the mask with the "Hard Contact" mode.

A posterior development step inside a beaker filled with MICROPOSIT™ MF-CD-26 developer dissolves the exposed areas of the resists by binding with the particles with broken bonds (Fig. 3.13d), exposing the underlying substrate. In this process, the sample is first immersed and constantly stirred in the developer for 80 s and cleaned in a DIW beaker for 90 s, then, such immersions are repeated for 10 and 15 s in developer and DIW, respectively. Lastly, the sample is dried with a nitrogen gun to ensure the proper removal of the solvents and the chemically altered resists.

To precisely etch the pattern on the exposed substrate areas, the sample is introduced into the vacuum chamber of an Oxford Instruments PlasmaPro® NGP80 reactive ion etcher (RIE), where an ion bombardment is directed towards the surface (Fig. 3.13e), selectively interacting with the substrate material due to a preestablished recipe. Such recipes are described in Table 3.1, where t_{etch} is the etching time, Φ_g is the reactive gas flux, $p_{chamber}$ is the maximum pressure inside the vacuum chamber during the etching, P_{RF} is the radiofrequency power that excites the plasma and V_{DC} is the bias between the anode and cathode of the system. The high energy and reactivity of the ions allow them to break and bind with the bonds of the exposed surfaces, etching a hole in the substrate whose thickness depends on the energy of the ions and the exposure time.

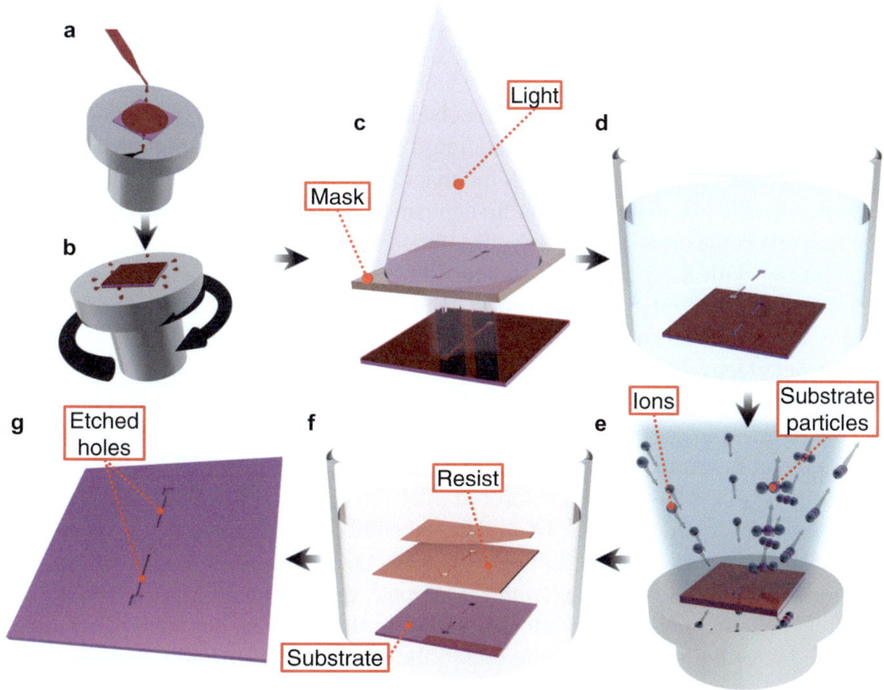

Fig. 3.13 Plasma etching of a substrate. **a** A liquid photoresist is dispensed with a pipette on the substrate, **b** the substrate is spun to flatten the photoresist, **c** optical radiation crosses through the pattern of a lithography mask to impinge on the photoresist, **d** the chemically modified region of the photoresist is removed from the substrate inside a beaker filled with developer, **e** plasma ions collide with the exposed parts of the substrate, reacting and sputtering with the topmost particles to etch the first layers of its surface, **f** the remaining photoresist is removed from the substrate inside a beaker filled with solvent and **g** the holes have been etched in the substrate

Table 3.1 Reactive ion etching recipes for different substrates

Substrate	t_{etch} (s)	Φ_g (SCCM)	$p_{chamber}$ (Pa)	P_{RF} (W)	V_{DC} (V)
SiO_2	150	25 (CHF_3) + 5 (Ar)	7.33	150	360
Si_3N_4	90	50 (CHF_3) + 5 (O_2)			355
Silicon	30	12 (CHF_3) + 30 (SF_6) + 10 (O_2)	1.33	80	100

The remaining photoresist layers are lifted-off from the substrate by following an additional solvent cleaning (Fig. 3.13f). In this process, the sample is first immersed in acetone for 1 h at RT. This removes the majority of the attached resists, however, a second immersion into IPA for 2 min, an ultrasonication in both acetone and IPA for 10 and 5 min, respectively, and a nitrogen drying ensure a proper cleanness of the patterned substrate (Fig. 3.13g).

3.4.2 Metallic Contacts

Metals are used as electrodes to input an electrical signal to the studied material and read the generated output. By shaping such electrodes, one can create multiple types of devices, e.g., transistors or probe stations. Thus, precise deposition processes are required to link the nanodevice with the macroworld.

In this thesis, two different metallic deposition procedures have been undertaken. The first one is the creation of electrodes directly on a standard substrate via thermal evaporation, with the possibility of depositing two-dimensional structures on top, creating a bottom contact configuration, while the second one consists of an indium needle deposition on top of a pre-exfoliated 2D flake or heterostructure to form a top contact electrode configuration. The wider parts of these electrodes are far from the target material and have millimetre-sized areas, hence, allowing their bonding to external wires, pins or chip carriers with silver paste or soldering.

The metal thermal evaporation is performed in the Lancaster University Physics Department QTC Class 1000 cleanroom after the optical lithography and development steps described in Sect. 3.4.1. The substrate/resist ensemble is fixed on the stage of a Moorfield Nanotechnology MiniLab 060 T60M thermal evaporator and positioned inside the vacuum chamber. Once the vacuum value of 2×10^{-3} Pa has been reached, the metal evaporation can start. In this process, a metallic thin film is deposited over the exposed sample due to the sputtering of metallic particles from an evaporation source inside a resistively heated ceramic basket (Fig. 3.14a). The inert nature of gold (preventing oxidation), its high affinity with surface chemical groups (increasing the easiness of deposition) and its good electronic performance position this metal as the material of choice for the electrodes. Nevertheless, a thinner titanium film is first deposited to increase the adhesion of gold with the substrate. The thickness of the thin films is monitored until it reaches approximately 2–3 nm of titanium and 20 nm of gold, given that, if the electrodes are too thick, the 2D materials' deposition yield is significantly decreased.

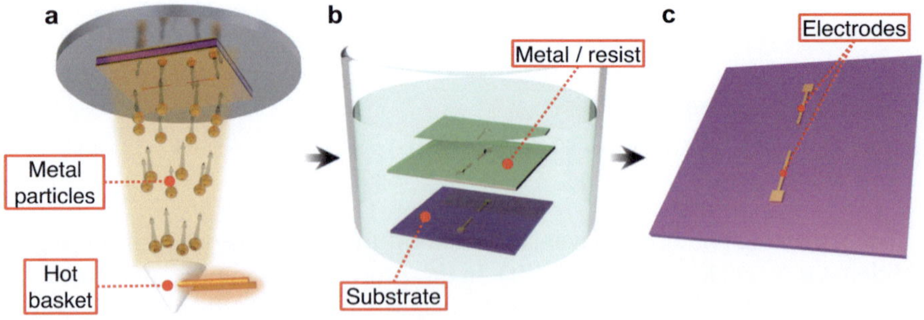

Fig. 3.14 Metal thermal evaporation on a substrate. **a** A substrate with a prepatterned photoresist is subjected to the deposition of metallic particles emanating from a hot basket filled with metal, **b** the contacting layers of deposited metal and photoresist are lifted-off from the substrate inside a beaker filled with remover and **c** the final electrodes have been deposited on the substrate

After the deposition, the excess photoresist and metallic layers are lifted-off by introducing the patterned structure into a beaker containing MICROPOSIT™ 1165 remover (Fig. 3.14b). This last step is divided into two subprocesses. In the first one, the sample is left for 30 min inside the liquid at a temperature of 90 °C and the lift-off is aided by gently blowing bubbles with a pipette towards the substrate. The second process comprises an immersion of the sample in a beaker with fresh remover for 5–15 min also at 90 °C to lift-off the remaining photoresist and metal. Lastly, the sample is cleaned in acetone and IPA and dried with a nitrogen gun, leaving the metallic electrodes ready for the 2D materials' deposition (Fig. 3.14c).

The deposition of indium electrodes is performed on a transfer station setup similar to that described in Sect. 3.3, with some modifications. The glass slide micromanipulator is replaced by a needle-arm micromanipulator, and the sample stage is upgraded to one with a larger surface capable of reaching higher temperatures (Fig. 3.15a). The micromanipulator can independently move the needle-arm in the three spatial directions, and it is mounted with an approximate tilted angle of 30° with respect to the horizontal plane for a more convenient indium deposition. A steel fine-apex needle is clamped on a plastic transparent arm, which, in turn, is attached to the micromanipulator. The stage, which can be moved manually, is bonded to the same TEC-RTD-resistor configuration utilised for the temperature monitoring during the heterostructure assembly. However, the resistors are switched to Bourns® PWR220T-35 10 Ω resistors, arranged in a 2-by-2 parallel scheme. The sample is secured to the stage with double-sided carbon tape, placing the solid indium bead as close as possible to it (Fig. 3.15b).

The stage temperature is increased enough to exceed indium's melting point (156.60 °C). Hence, the indium bead becomes liquid but maintains its drop-like shape thanks to its high surface tension [11], allowing the metallic needle to be inserted (Fig. 3.15c). The needle is then slowly retracted, pulling indium out from the bead (Fig. 3.15d), and, as the pulled indium is separated from the liquid bead and the hot stage, it solidifies forming a sharp indium needle (Fig. 3.15e). The indium needle's length and sharpness depend on the bead's temperature and the retraction speed and continuity. For optimal sharpness, the retraction is performed at 160 °C at a low, steady speed. If the indium needle is too long, its stability decreases, causing the tip of the needle to oscillate and complicating the following deposition.

To deposit the indium needle onto the sample, the area of interest is located on the sample's surface via the optical microscope and centred with the indium needle's apex. The stage's temperature is then increased to 165–170 °C to ensure that the sample's top surface is hot enough to melt the indium, since it may not be at the same temperature as the stage due to the substrate's thickness and composition. The needle is lowered slowly with fine adjustments until it contacts the area of interest with the apex (Fig. 3.15f). It is continued lowering until the desired length and amount of indium is deposited, forming the electrode with expanded width away from the apex. Finally, the arm is quickly retracted to break contact, leaving the indium needle attached to the sample (Fig. 3.15g).

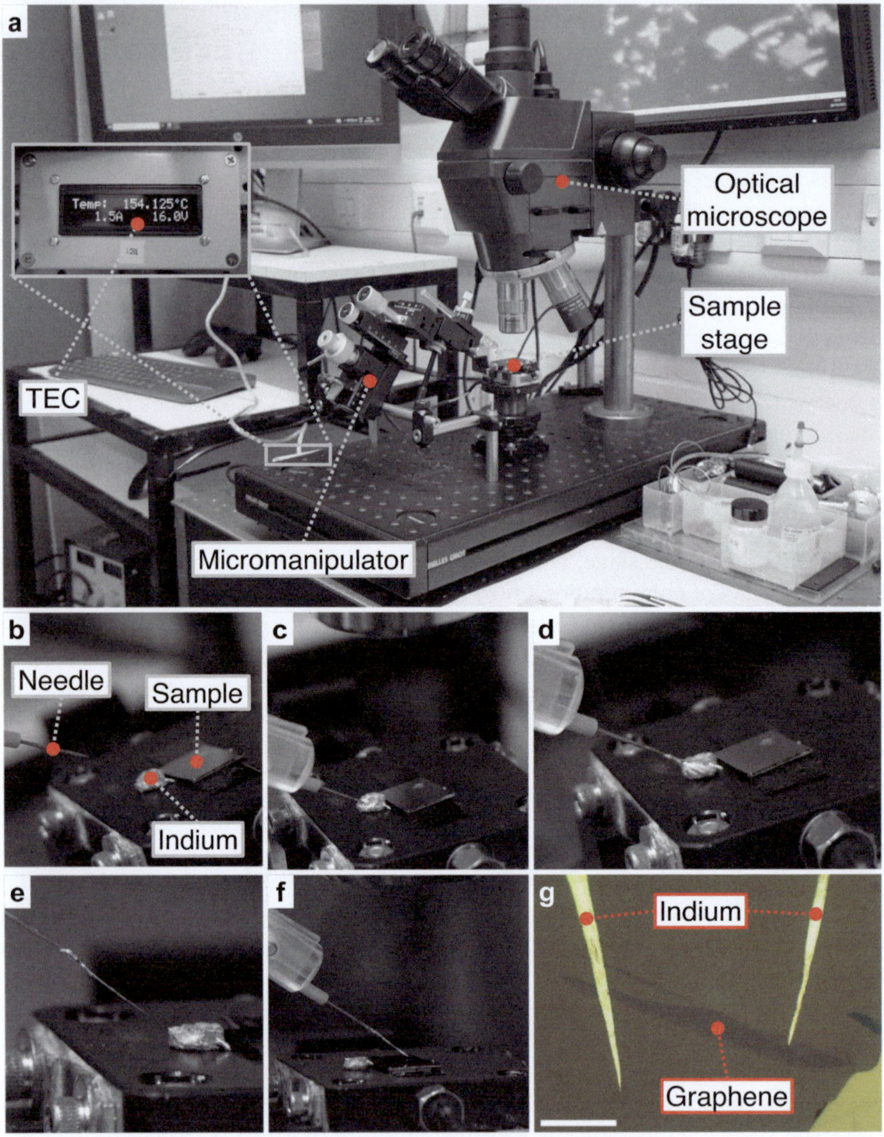

Fig. 3.15 Fabrication of indium electrodes. **a** Employed setup with the main parts highlighted (TEC stands for temperature controller), **b** close-up look of the sample stage with the main parts highlighted, **c** needle dipping into the liquid indium bead, **d** retracting needle pulling the liquid indium, **e** fully retracted indium needle, **f** indium needle positioned over the sample and **g** optical microscope image of a graphene flake with the deposited indium electrodes (scale bar is 40 μm)

3.5 Beam Exit Cross-Sectional Polishing

Most nanocharacterisation techniques are only capable of obtaining information from the sample's surface up to a few nm of the inner structure. To access the deeper buried layers, focused ion beam (FIB) and argon ion sectioning are currently the most common milling techniques, however, their perpendicular ion beams impinging on the sample surface damage its topmost layers and, in the case of FIB, implant gallium atoms on the exposed surface, affecting its physical properties [12, 13]. Beam exit cross-sectional polishing (BEXP) is a precise nano-sectioning tool that uses non-reactive argon ion beams impinging on the side of the sample at a negative glancing angle and exiting through the sample's top surface [14]. The shallow angle (~ 3–5°) between the beams and the sample creates a damage-free wedge-like oblique section easily accessible for SPM studies (Fig. 3.16c). Moreover, the inert nature of argon minimises the resulting roughness of the cut and the probability of particle deposition, preserving the original physical characteristics of the wedged material [12, 13, 15].

The process is conducted inside the vacuum-sealed chamber of a Leica EM TIC 3X ion beam milling system. The sample is fixed on a tilted stage at 3 or 5° with respect to the horizontal plane, and an etch-resistant mask is calibrated to cover most of the sample's side, only exposing the top area to be milled. In the case where a flake situated near the substrate's edge is the point of interest (Fig. 3.16d), the position of such flake is also centred accordingly to the ion beams' convergent point. Then, the vacuum chamber is sealed and pumped down to ~ 2×10^{-3} Pa and the argon lines are purged of contaminants.

When the sectioning operation is started, three ionic guns generate three incident overlapping coplanar beams (Fig. 3.16a). Each gun is composed of a rear and frontal cathode and a central anode, generating an electric field that ionises the pressurised argon gas within the gun enclosure. The path of the plasma beam is only allowed through an aperture at the frontal cathode from where the argon ions exit the gun and accelerate towards the exposed sample and mask (Fig. 3.16b). To enhance the quality of the cut and decrease the wear of the BEXP system, the energy of the beams is engineered in different steps. First, two consecutive warm-up steps activate and prepare the guns at 1 kV, 1 mA for 1 min and at 3 kV, 1 mA for 2 min, respectively. Then, the beams' energy is increased up to the milling parameters, specifically 7 kV, 2.60 mA for 7 h. The 7 h of sectioning are sufficient for the beams to fully cut through the exposed area and exit via the top sample surface, hence the "beam exit" name. This results in the wedge-like section of the subsurface layers of both the substrate and the top thin film or flake (Fig. 3.16e). Finally, a two-step post-polishing process is conducted to ensure a sub-nm near-atomic roughness of the wedge cut (Fig. 3.16f) at 3 kV, 1 mA for 10 min and 1 kV, 1 mA for 30 min.

Thanks to the grazing incident angle of the beams, BEXP results in 0.50–1 mm wide wedge areas from a minimum depth of cut (20–40 μm), expanding the inner sample features for a precise characterisation even of nano-thin layers. For instance, a 2 nm thin layer can be expanded to 10–12 nm of wedged width, enhancing the

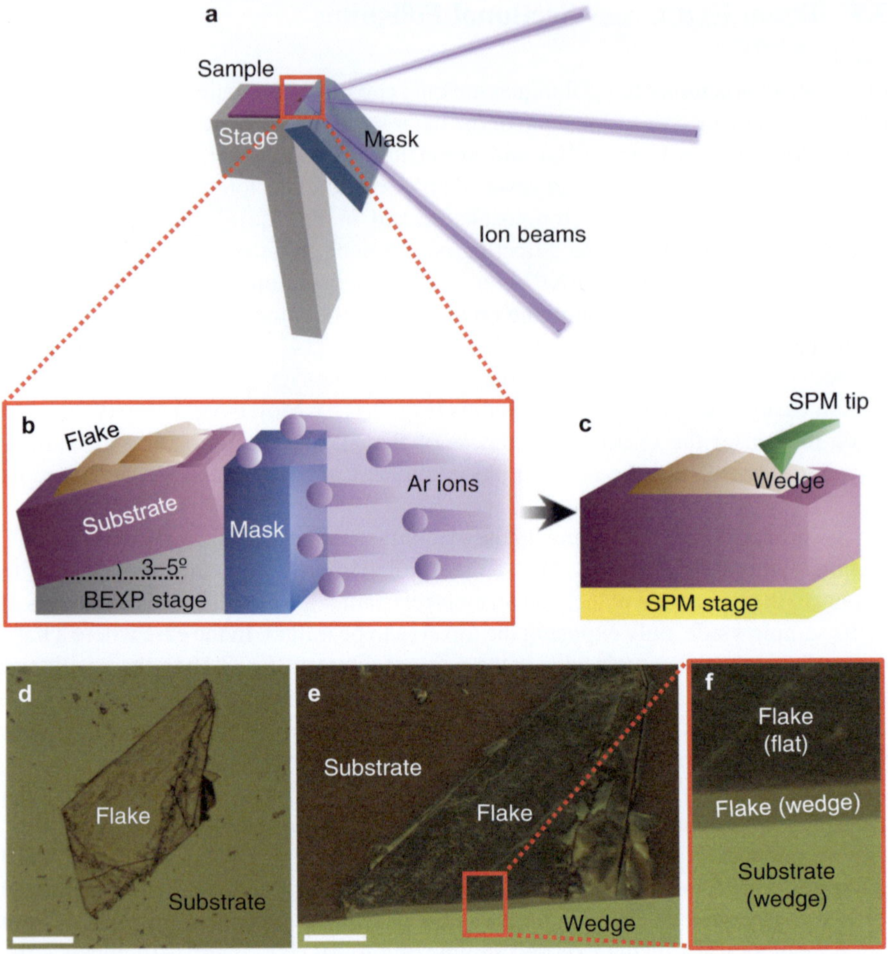

Fig. 3.16 **a** Scheme of the basic BEXP components, **b** magnified view of the cut operation, **c** scheme of the prepared sample after the sectioning, graphite flake **d** before and **e** after the BEXP process (scale bars are 60 and 30 μm, respectively) and **f** magnified view of the wedge and flat areas

SPM thickness measures. Combining BEXP and SPM is faster, more cost-effective and can measure bigger areas than other tools, such as TEM.

In this project, BEXP is applied along with high sensitivity SPM techniques to obtain a gradient of thickness related to other mechanical, thermal and electronic nanoscale properties of the measured samples in situ, meaning that these properties can be related and monitored in a single image as the thickness periodically varies from the thin, e.g., monolayer, to the thicker (hundreds of nm) layers of the material.

3.6 Summary

The multiple procedures of sample fabrication employed in this thesis have been described in detail in the present chapter. Specifically, the fundamentals of classic mechanical exfoliation have been discussed for each of the four most utilised 2D materials, i.e., graphene, h-BN, MoS_2 and InSe, highlighting the importance of prior treatment of the supporting substrate. Following, a time- and cost-effective methodology for the stacking of such materials forming homo- or heterostructures has been summarised, having also tested other processes that lacked such convenience. The synthesis of nanoscale devices based on 2D materials has been subsequently described, focusing on the geometrical etching or deposition of the involved materials in the shape of holes or electrodes, respectively. To finalise, BEXP has been proposed as the alternative to FIB for accessing the inner structure of vdW samples, with the added benefits of expanding the measurable areas of such layers with minimal surface damage.

All the synthesis techniques herein described are mostly relevant for research purposes, meaning that they focus on producing a small number of adequate quality samples intended to be analysed. These methods have to be highly tunable to adapt to the different desired resulting properties, hence, they generally consist of bespoke steps that provide versatile and selective capabilities. However, these processes are not as time-, quality- and cost-effective as their industrial counterparts, which prioritise such aspects since they are designed for batch assembly and distribution.

References

1. Blake, P., et al. (2007). Making graphene visible. *Applied Physics Letters, 91*, 063124.
2. Hattori, Y., Taniguchi, T., Watanabe, K., & Kitamura, M. (2022). Enhancement of the contrast for a hexagonal boron nitride monolayer placed on a silicon nitride/silicon substrate. *Applied Physics Express, 15*, 086502.
3. Rodica, G., et al. (2011). Study of SiO_2/Si interface by surface techniques. In B. Sukumar (ed) *Crystalline silicon* (Chap. 2, pp. 23–42). IntechOpen
4. Qin, B., et al. (2020). Substrates in the synthesis of two-dimensional materials via chemical vapor deposition. *Chemistry of Materials, 32*, 10321–10347.
5. Gorbachev, R. V., et al. (2011). Hunting for monolayer boron nitride: Optical and raman signatures. *Small (Weinheim an der Bergstrasse, Germany), 7*, 465–468.
6. Ling, F., et al. (2018). Optimizing edges and defects of supported MoS_2 catalysts for hydrogen evolution via an external electric field. *Physical Chemistry Chemical Physics: PCCP, 20*, 26083–26090.
7. Onodera, M., Masubuchi, S., Moriya, R., & Machida, T. (2020). Assembly of van der Waals heterostructures: Exfoliation, searching, and stacking of 2D materials. *Japanese Journal of Applied Physics, 59*, 010101.
8. Pizzocchero, F., et al. (2016). The hot pick-up technique for batch assembly of van der Waals heterostructures. *Nature Communications, 7*, 11894.
9. Kinoshita, K., et al. (2019). Dry release transfer of graphene and few-layer h-BN by utilizing thermoplasticity of polypropylene carbonate. *NPJ 2D Materials and Application, 3*, 22.
10. Castellanos-Gomez, A., et al. (2014). Deterministic transfer of two-dimensional materials by all-dry viscoelastic stamping. *2D Materials, 1,* 011002.

11. Alchagirov, B. B., Dadashev, R. K., Dyshekova, F. F., & Elimkhanov, D. Z. (2014). The surface tension of indium: Methods and results of investigations. *High Temperature, 52*, 920–938.
12. Kolosov, O. V., Grishin, I., & Jones, R. (2011). Material sensitive scanning probe microscopy of subsurface semiconductor nanostructures via beam exit Ar ion polishing. *Nanotechnology, 22*, 185702.
13. Robson, A. J., et al. (2013). High-accuracy analysis of nanoscale semiconductor layers using beam-exit Ar-ion polishing and scanning probe microscopy. *ACS Applied Materials & Interfaces, 5*, 3241–3245.
14. Kolosov, O. V., & Grishin, I. (2012). Method and apparatus for ion beam polishing. United States patent US 9082587 B2.
15. Bosse, J. L., Grishin, I., Huey, B. D., & Kolosov, O. V. (2014). Nanomechanical morphology of amorphous, transition, and crystalline domains in phase change memory thin films. *Applied Surface Science, 314*, 151–157.

Chapter 4
Thermal Transport in Two-Dimensional Nanolayers

Abstract In this chapter, the heat transport in two-dimensional (2D) nanolayers is investigated, with a focus on achieving directional control and understanding anisotropic thermal behaviour at the nanoscale. The chapter introduces a novel characterisation technique, known as cross-sectional scanning thermal microscopy (xSThM), used to probe in-plane and cross-plane thermal conductivities in layered semiconductor materials with high spatial resolution. This method enables the direct measurement of anisotropic thermal transport, addressing limitations of conventional approaches and allowing for quantitative evaluation of thermal conductivity components in complex 2D systems. Following the experimental measurements, an analytical model is developed to describe heat flow in anisotropic media, forming the theoretical basis for interpreting the obtained data. To further validate the results and extend the analysis, finite element simulations are conducted, replicating experimental conditions and providing insight into thermal gradients and interfacial effects. The chapter also includes a systematic error evaluation using an isotropic reference structure to assess the reliability and sensitivity of the measurements. In addition to extracting directional thermal conductivities, the methodology enables estimation of the upper bounds of the interfacial thermal resistance between 2D materials and their substrates. Together, these results offer a robust platform for analysing thermal transport in layered nanostructures, with implications for future thermoelectric and heat management applications.

4.1 Introduction

Enhancing the efficiency of modern technology requires the engineering of its energy conversion processes. Specifically, electronic systems are intrinsically related to thermal energy, where, depending on the application, heat transport has to be obstructed, directed or maximised, e.g., in thermoelectric devices or heat spreaders in processor chips [1–3]. 2D materials have become a useful tool in this regard since their properties can be easily adapted to the desired thermal outcome [4, 5]. For

S. Gonzalez-Munoz, *Interplay of Nanoscale Physical Properties in 2D Materials and Heterostructures*, Springer Theses, https://doi.org/10.1007/978-3-032-14271-9_4

instance, tuning the thickness or composition of 2D semiconductors allows to engineer a device with superior TE characteristics [6, 7], given that their band structures provide high electrical conductivity and Seebeck coefficient while restricting the phonon-dominated thermal conductivity [8].

InSe and 2D lead iodide perovskites have been recently catalogued as suitable TE candidates thanks to their direct bandgap and high electrical conductivity [9, 10]. However, the fundamental anisotropic thermal transport characterisation in such 2D systems is limited to theoretical data or non-trivial methodologies, with experimental findings mostly focusing on the electronic behaviour, 3D heat transport or disregarding their anisotropic thermal properties [11–13].

SThM employs a nanoscale tip sensitive to heat flow, being able to measure non-destructively the local thermal properties of atomically layered materials with an optimal spatial resolution. Nevertheless, the current SThM standalone capabilities are not able to deconvolute the heat flow directionality in anisotropic materials, nor to differentiate between the thermal transport in the layered material and the substrate as well as in the tip/sample and sample/substrate interfaces [14, 15]. Each of these subsystems plays a vital role in the overall SThM thermal response, being influenced by, for instance, phonon scattering at the edges or interfaces. Hence, a methodology capable of individually determining the intrinsic thermal parameters accounting also for their directionality is required.

In this chapter, a novel functional approach called cross-sectional scanning thermal microscopy (xSThM) is used to address said limitations [15] By combining BEXP and SThM, one can map the inner topographical and thermal structure of different 2D materials supported on silicon and SiO_2 substrates in a single scan. Additionally, the absolute quantification of the thermal properties of the layered materials along their different directions and their corresponding interfaces is achieved with an experimental data analysis via a "Muzychka-Spièce" model [15–19] and validated by FEA simulations and xSThM proof-of-concept of an established isotropic material. Three different anisotropic 2D materials are studied herein. γ-InSe and phenethylammonium lead iodide [$(PEA)_2PbI_4$] perovskite are prospective TE materials with expected low thermal conductivities that combine with superior electronic behaviour. Graphene, on the other hand, has a major anisotropic ratio between its elevated in-plane and moderately low cross-plane thermal conductivities, hence, it is an interesting reference to the lower heat conductivities of the two aforementioned 2D semiconductors.

The contribution of Dr. Abhishek Maiti is acknowledged since he performed the perovskite sample fabrication as well as part of its SThM characterisation and interpretation. The fabrication, characterisation and analysis of the rest of the samples were performed by myself in the Lancaster University Physics Department, as well as the perovskite data analysis. The results reported in this chapter have been published in Gonzalez-Munoz, S. et al. *Adv. Mater. Interfaces* **10** (2023) and Maiti, A. et al. *Phys. Rev. Mater.* **7** (2023) (see "List of Publications").

4.2 Cross-Sectional Scanning Thermal Microscopy

Sample Fabrication

The graphite sample is fabricated by dry exfoliation (see Sect. 3.2.2) from a ProGraphite Natural Graphite Giant Flake bulk crystal and posterior deposition on a pretreated silicon substrate. The flake of interest is chosen to be close to one of the substrate edges and to have a thickness of approximately 400 nm. The bulk crystal used in the exfoliation of the γ-InSe sample is fabricated by Prof. Zakhar D. Kovalyuk and Dr. Zakhar R. Kudrynskyi at the National Academy of Sciences of Ukraine using the Bridgman method from a polycrystalline melt of $In_{1.03}Se_{0.97}$. The target flake is mechanically exfoliated on a silicon substrate (see Sect. 3.2.3) close to the substrate's edge and has a thickness of ~ 100 nm.

The perovskite sample consists of a thin film coated on a 300 nm SiO_2/silicon substrate. The precursor material forming the perovskite thin film includes phenethylammonium iodide (PEAI, 99%), lead iodide (PbI_2, 99%), and anhydrous N,N-dimethylformamide (DMF, 99.80%), purchased from Sigma-Aldrich®. A homogeneous solution of such materials is formed by dissolving 2 mol of PEAI as well as 1 mol of PbI_2 into DMF and kept under continuous stirring at 70 °C. The solution is spin coated on the previously treated substrate at 2500 rpm for 30 s and annealed at 80 °C for 15 min. To access the internal layers of the three 2D samples and achieve a near-atomically smooth tilted wedge surface of continuously varied thickness values in a single sample ensemble, BEXP is employed (see Sect. 3.5).

High Vacuum Scanning Thermal Microscopy Setup and Calibration

Topographical and thermal characterisation of the cross-sectioned samples is conducted via a high vacuum (HV) NT-MDT® Spectrum Instruments® Smena SThM system with custom-built electronics (Fig. 4.1a), rapidly transferring the samples to the vacuum chamber to avoid their exposure to ambient conditions and the formation of oxide layers on the BEXP surfaces.

The model of the employed probes is a KNT-SThM-2an from Kelvin Nanotechnology, with a spring constant of 0.40 Nm^{-1} and a tip radius < 100 nm. Such type of microfabricated Si_3N_4 cantilever and gold pad probes have incorporated a palladium resistive heater close to the tip apex, electrically connected to a balanced BRG receiving 3 V DC and 4 V peak-to-peak AC excitation at a frequency of 91 kHz from a Keithley 3390 arbitrary waveform generator (GEN), and supplying 0.30 V DC and 0.40 V AC signal to the probe. The received power produces nearly constant Joule heating in the tip apex, however, when the tip is brought into contact with the sample, heat additionally dissipates through the tip/sample contact interface (usually of lateral dimensions of a few tens of nm) towards the layered material/substrate ensemble according to its thermal properties, modifying the probe's temperature and the heater's electrical resistance [20]. This resistance variation is converted to voltage to represent the xSThM thermal signal. The output thermal signal is filtered in a Stanford Research Systems SR650 dual channel filter (FLT), measured with a Stanford Research Systems SR830 DSP LIA and sent to the SPM controller. The signal

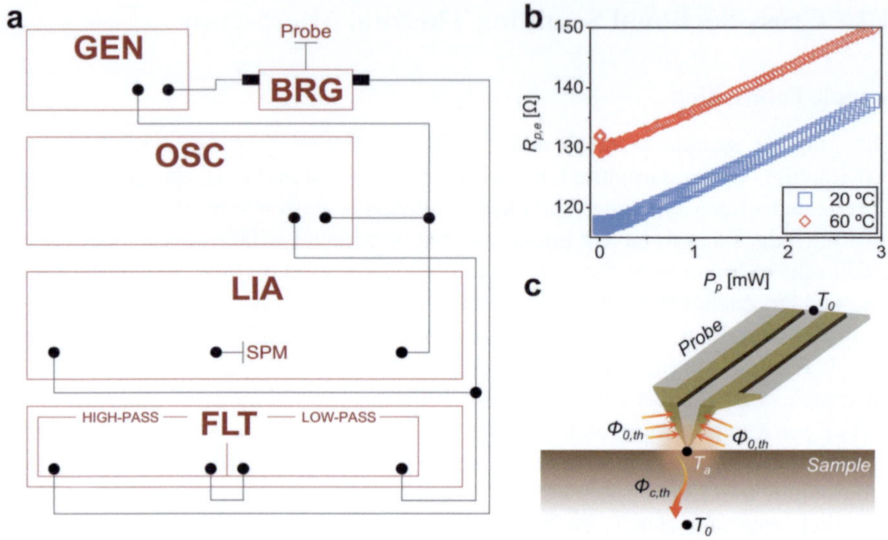

Fig. 4.1 **a** SThM custom-built electronics scheme (GEN, BRG, OSC, LIA and FLT are the generator, bridge, oscilloscope, lock-in amplifier and filter, respectively), **b** typical probe's electrical resistance versus received power plots at different temperatures and **c** SThM probe/sample heat flux (Φ) scheme

can also be qualitatively monitored with a Rigol DS1074Z oscilloscope (OSC). This results in simultaneous topography and xSThM thermal signal maps, where the latter is proportional to the probe temperature rise with respect to the ambient temperature. The thermal decrease qualitatively corresponds to the increase in the tip/sample heat flow.

In order to minimise heat transfer through undesired channels, the measurements are performed under HV ($< 2 \times 10^{-6}$ mbar) and RT conditions (300.50 ± 0.20 K), monitoring the sample stage temperature with an RTD sensor (RS PRO PT1000). To minimise external vibrations, the stage is suspended with springs and stabilised by magnetic eddy current dampers. The HV environment prevents conduction and convection parasitic heat losses, which could originate, e.g., in air measurements or through a water meniscus, given that the only media in contact are the tip apex and the analysed material. Also, the low relative temperature involved in xSThM minimises the radiation heat losses proportional to the 3rd power of the temperature [21]. The scans are performed at a minimal indentation setpoint to avoid any mechanical effect on the thermal signal. Also, each of the materials is measured by a different probe from the same KNT-SThM-2an batch, avoiding influence from a previously used tip.

Prior calibration of the probes is needed in order to convert the xSThM thermal signal to the quantitative heat properties of the samples. Specifically, the probe thermal resistance values ($R_{p,th}$) under HV conditions must be obtained. To do so, the probes are mounted on a Peltier module driven by a TEC and their overall electrical resistances at 20 and 60 °C are measured while the probes' current is varied from minimal values to 1.50 mA. The employed probes contain two 100 Ω resistors in series on the cantilever, hence, a total value of 200 Ω must be subtracted from the obtained probe's overall electrical resistance to determine the corrected palladium heater resistance value ($R_{p,e}$). These values can be compared to the power applied on the probes (P_p) (Fig. 4.1b) and extrapolated to zero power in order to obtain the probe thermal resistance under ambient conditions via the following interpolation equation (see details in Sect. A.2.3).

$$\left(R_{p,th}\right)_{amb} = \frac{40\left\{\left[R_{p,e}(P_p)\right]_{20\,^\circ\mathrm{C}} - \left[R_{p,e}(0)\right]_{20\,^\circ\mathrm{C}}\right\}}{\left\{\left[R_{p,e}(0)\right]_{60\,^\circ\mathrm{C}} - \left[R_{p,e}(0)\right]_{20\,^\circ\mathrm{C}}\right\}\left(P_p\right)_{20\,^\circ\mathrm{C}}} \tag{4.1}$$

The linearity in the relation between the thermal signal and the temperature of the probe as well as the consistency of the probe's power can be exploited to determine the $R_{p,th}$ values under high vacuum conditions. Hence, the ratio between the non-contact thermal signals (V_{nc}) under ambient and vacuum conditions can be used to obtain the HV $R_{p,th}$ as:

$$\left(R_{p,th}\right)_{HV} = \frac{\left(R_{p,th}\right)_{amb}\left(V_{nc}\right)_{HV}}{\left(V_{nc}\right)_{amb}} \tag{4.2}$$

where the V_{nc} values are obtained by averaging the non-contact voltages recorded in multiple approach and retract curves under both conditions.

In spite of the heat being homogeneously supplied over the entirety of the resistive heater, the narrow triangular geometry of the tip increases the apex temperatures. Also, when the tip and sample are in contact, a temperature gradient is heterogeneously distributed given the created heat flow constricted path, with the temperature increase measured by the resistive sensor [16]. To account for this spatial distribution of the temperature, a sensitivity coefficient (α_c) is reported for the first time in SThM measurements as a correction factor for the experimental xSThM data, as introduced in the following sections. However, it is still essential to understand the quantitative relation between the tip apex temperature and its average distribution, affecting the nature of α_c, with the intrinsic in-contact heat transport channels depicted in Fig. 4.1c. Since the heat transport equations have a linear nature, one can relate the different spatial probe temperatures with the intrinsically generated ($\Phi_{0,th}$) and the through-the-contact ($\Phi_{c,th}$) heat fluxes as:

$$T_p = R_{p,th,eff}\,\Phi_{0,th} - R_{c,eff}\,\Phi_{c,th} \tag{4.3}$$

$$T_a - T_0 = R_{p,th}\Phi_{0,th} - R_c\Phi_{c,th} \tag{4.4}$$

where T_p is the average probe temperature and T_a and T_0 are the apex and ambient temperatures, respectively. $R_{p,th,eff}$ and $R_{c,eff}$ are the probe and contact effective thermal resistances obtained via modelling, whereas R_c is the tip/sample thermal resistance obtained in the following sections. Given that $R_{p,th}$ and $\Phi_{0,th}$ can be obtained experimentally, the only two remaining unknowns are T_a and $\Phi_{c,th}$. Hence, by applying Eq. (4.5), T_a can be obtained and compared to T_p to find the temperature relation sensitivity coefficient ($\alpha_c = T_a T_p^{-1}$).

$$T_a = R_{p,th}\Phi_{0,th} - R_c \frac{R_{p,th,eff}\Phi_{0,th} - T_p}{R_{c,eff}} + T_0 \tag{4.5}$$

Hence, this geometric coefficient can be used to derive the different thermal signal responses of the probes to the same amount of heat flux present during their non-contact self-heating and the heat flux towards the samples while in-contact. This parameter is exclusively dependent on the tip apex and heater geometry, therefore, it is not influenced by the thermal properties of the measured samples or the heat applied to the probes [16].

Topography and Thermal Response Measurements

A preliminary study of a γ-InSe flake with different thicknesses on a flat silicon substrate is performed, following the methodology explained in a previous study [11]. Figure 4.2a andc show its topographic map and extracted profile, respectively, differentiating between the flake's thicknesses (h) depending on the tip position (x). However, when the thermal map and its corresponding profile (Figure 4.2b and d, respectively) are analysed, an approximately constant thermal signal (V_x) is obtained without the expected thermal signal versus thickness dependent transition. This nearly constant behaviour is due to the elevated tip/material interfacial thermal resistance (R_c), which is independent of the thickness [17] and obscures the much lower thickness-dependent spreading thermal resistance (R_S) values that represent the sample's heat transport properties. For this reason, a methodology capable of simultaneously recording the discrete variation of the sample's thermal signal versus its thickness and eliminating the influence of R_c is required. xSThM opens a pathway for the former possibility, given that the tip can access and record the corresponding thermal signal of each individual thickness value thanks to the BEXP cut while keeping the R_c constant, allowing to quantify the heat transport features in the anisotropic samples.

xSThM studies are conducted on the three samples previously fabricated, where the 2D material/substrate ensemble is analysed. For the subsequent fitting of the obtained SThM results, the tip/sample contact radius (a_c) must be measured. This is done by also performing xSThM on isotropic SiO_2/silicon reference layers in adjacent samples on the same stage in the case of the γ-InSe/silicon and graphite/silicon structures or directly in the same perovskite/SiO_2/silicon sample. All these isotropic ensembles are composed of a 300 nm thermally grown SiO_2 layer ($\kappa = 1.40 \pm 0.20$ W m^{-1} K^{-1}) on a silicon substrate ($\kappa_s = 130 \pm 20$ W m^{-1} K^{-1}) and are cut following the standard BEXP procedure [22–25].

Figure 4.3 shows the 1 μm^2 xSThM topographical and thermal signal maps for each of the anisotropic samples (Fig. 4.3 g–l). To simplify the discussion, the isotropic reference samples corresponding to the γ-InSe/silicon, graphite/silicon and perovskite/SiO_2 structures are named RS1 Fig. 4.3a and d), RS2 (Fig. 4.3b and e) and

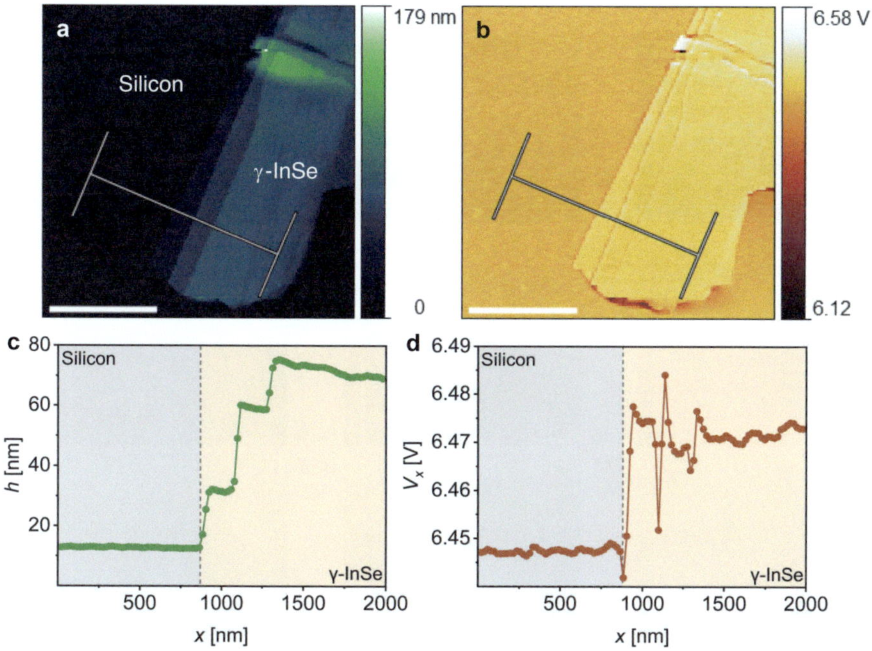

Fig. 4.2 SThM of γ-InSe flake on a flat silicon substrate. **a** Topography and **b** thermal signal maps (scale bars are 1 μm). **c** Topography and **d** thermal signal profiles extracted from **a** and **b**, respectively

RS3 (Fig. 4.3c and f), respectively. Also, Fig. 4.4 represents the extracted profile data from the lines drawn in Fig. 4.3. Minimal surface roughness is observed along the topographic green profiles for all the samples, however, due to the different stability of the materials with respect to etching, a slight change occurs in the linear slope ($7.13 \pm 5.44\%$) between the substrates and the top layers. Nevertheless, the RMS roughness (ψ) and layer thickness (h) of the top materials can be quantified directly from the topography maps (see Table 4.1). The different roughness values between the anisotropic and isotropic samples may affect their tip/surface thermal resistance if the contact dimensions are in their same range, however, these interfacial sizes are typically 20–50 nm of radius [15], hence, one can safely assume that, given the bigger contact scale, a_c is identical in both structure types. Furthermore, the low roughness values validate the minimal damage caused by the BEXP, allowing to disregard its effect on the materials' nanoscale properties [15, 26, 27].

The thermal signal (V_x) versus position (x) profiles extracted from Fig. 4.3d–f and j–l are also shown in Fig. 4.4, where the boundaries between the top layers and substrates as well as the gradual increase of V_x proportional to the thickness can be distinctly identified. Qualitatively, the materials with higher thermal resistance can be related to an increased thermal signal in the same map, therefore, it is possible to state that the silicon layers are better overall thermal conductors than the SiO_2,

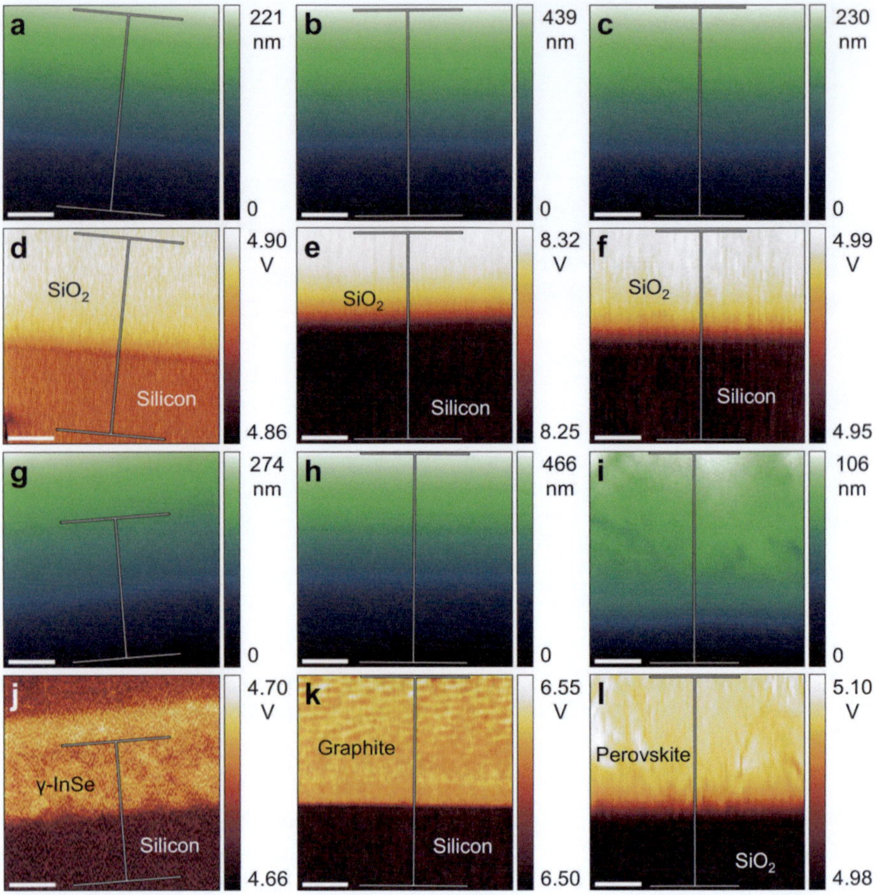

Fig. 4.3 xSThM maps of the samples. **a–c** Topographic and **d–f** thermal signal maps of the RS1, RS2 and RS3, respectively. **g–i** Topographic and **j–l** thermal signal maps of the corresponding γ-InSe/silicon, graphite/silicon and perovskite/SiO_2 samples, respectively. The drawn lines determine the extracted profiles shown in Fig. 4.4. Scale bars are 200 nm

γ-InSe and graphite and that perovskite is more thermally resistive than its SiO_2/silicon substrate. In the isotropic samples, the thermal signal (V_x) dependence on the thickness (h) is determined by the contact dimension (a_c). First, V_x linearly increases at thicknesses lower than the expected contact radius, then, it saturates at $h \gg a_c$. This thermal signal is converted to thermal resistance (R_x) with respect to h via Eq. (4.6) and the obtained values are employed as the experimental data to be fitted in the next section, determining the thermal parameters of the samples [15–17].

$$R_x(h) = \alpha_c \frac{V_x(h) R_{p,th}}{V_{nc} - V_x(h)} \tag{4.6}$$

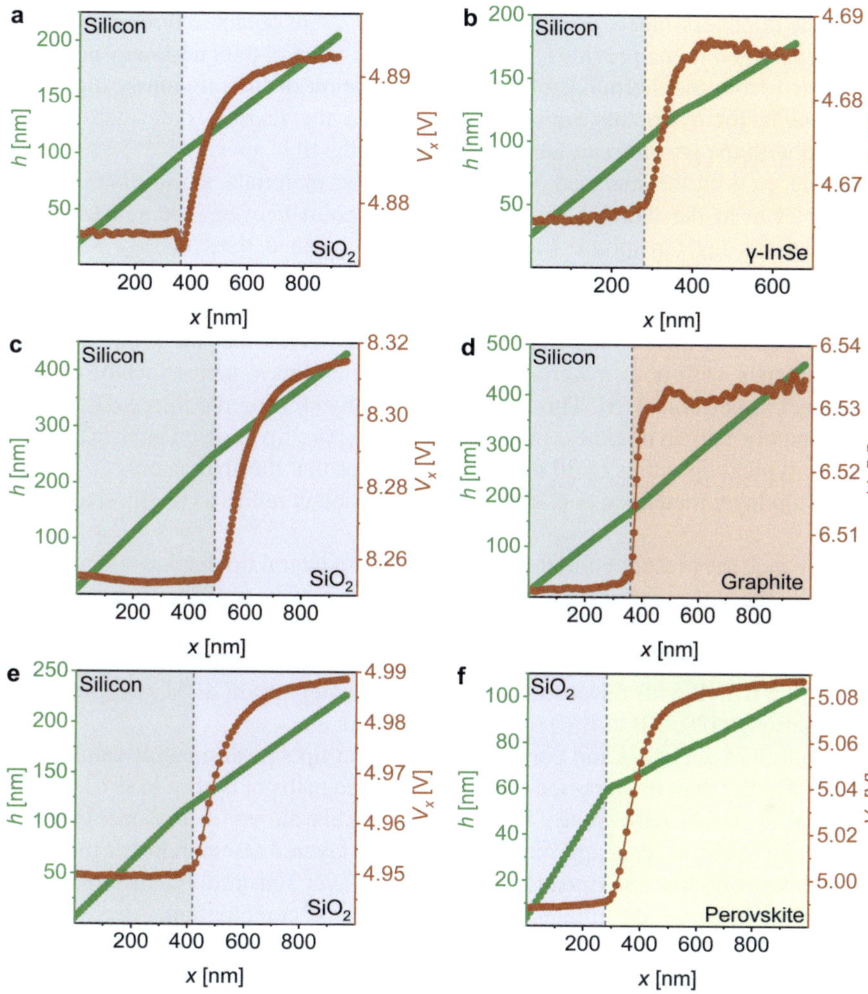

Fig. 4.4 Profiles extracted from the xSThM maps of Fig. 4.3. **a**, **c** and **e** Topography (green) and thermal signal (brown) profiles of the RS1, RS2 and RS3, respectively. **b**, **d** and **f** Topography (green) and thermal signal (brown) profiles of the corresponding γ-InSe/silicon, graphite/silicon and perovskite/SiO_2 samples, respectively

Table 4.1 Topographical parameters of the samples' top layers

	SiO_2 (RS1)	SiO_2 (RS2)	SiO_2 (RS3)	γ-InSe	Graphite	Perovskite
h (nm)	288.64	280.24	282.10	90.31	350.56	337.82
ψ (nm)	1.97	0.64	1.09	2.58	1.47	3.23

Where the thickness-independent $R_{p,th}$ as well as V_{nc} are obtained through the calibration explained in the previous section and the in-contact thickness-dependent V_x is extracted from the thermal profiles. The application of the previously discussed correction factor α_c in this equation ensures that the thermal data measured is weighted with the geometrical temperature rise in the tip's apex.

In order to limit the analysed data to the top layer materials, the initial thickness value is taken at the thermal signal intersection point between the substrate and the sloped top layer response (depicted by black dashed lines in Fig. 4.4), also corresponding to the topographical slope change. Nevertheless, the thermal signal data at the narrow transition zone is affected by the top layer/substrate interface (clearly visible as a dip, for instance, in Fig. 4.4a), given that the probe contacts simultaneously with both materials having different slopes, hence locally varying the contact dimensions [15]. This zone is avoided by shifting the initial data points approximately 6 nm in thickness (h_0) (corresponding to a tip position greater than the averaged typical tip radius of 30 nm), which ensures that the apex is only in contact with the top layer materials, conserving the same contact radius as the tip scans over them.

To enhance the present 6 nm thickness and 30 nm lateral tip displacement spatial resolution, the tip/sample contact radius would have to be smaller. This could be achieved by, e.g., reducing the tip apex radius incorporating a CNT at the end of the thermal probes [28] and/or performing a gentler post-polishing on the wedge surfaces either in the BEXP with fewer millivolts in the last step or via SPM sweeping with a stiff cantilever [29, 30].

The minimal thickness and corresponding lateral tip's position shift values used herein are larger than the reported phonon mean free paths of both γ-InSe (2.65 nm) [9] and lead iodide perovskite (4.30 nm) [31]. This allows to reasonably use a diffusion thermal transport approximation for both layered semiconductor materials while considering that the tip/surface R_c and top layer/substrate r_{int} are thickness-independent and are only influenced by the materials' combination. Also, one can assume a constant behaviour of the phonon scattering and electron/phonon interaction for layers thicker than the mean free path of the analysed 2D structures. The reported graphene mean free path ranges from 10 nm to 1 μm [32], hence, a ballistic approximation would be more accurate in the determination of its thermal properties. However, the large thickness of the studied flake and the expected high thermal conductivity can increase the number of scattering sources, reducing the mean free path and allowing the use of the diffusive approximation analytical method. Although this diffusive heat transport limitation impedes the thermal study of layers with thicknesses lower than the heat carrier's mean free path or lateral in-plane dimensions smaller than 1–2 μm [11], requiring modelling capable of accounting for the ballistic transport in atomically thin structures (e.g., the Boltzmann transport equation) [33], xSThM remains sensitive to the anisotropic thermal conductivity of the measured materials, as demonstrated below.

Given the independent contribution of the different elements present in the xSThM ensembles, the deconvolution of their thermal properties must be performed from the overall SThM thermal signal, in terms of thermal resistances. Figure 4.5 shows

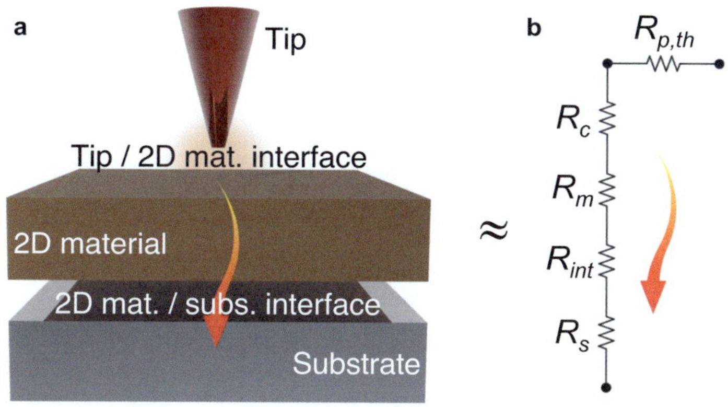

Fig. 4.5 Scheme of the thermal transport SThM model with an orange arrow indicating the heat flux's direction. **a** Tip/sample assembled layers and **b** equivalent thermal resistances for each internal and surrounding element (connected in series) and thermal resistance of the probe to its base (connected in parallel)

such thermal resistances connected in series, allowing for their representation as:

$$R_x = R_S + R_c = R_m + R_{int} + R_s + R_c \tag{4.7}$$

where R_x is the previously defined experimental xSThM thermal resistance that includes the spreading thermal resistance in the sample (R_S) as well as the tip/sample contact thermal resistance (R_c). R_S entails the top layer material, the material/substrate interface and the substrate thermal resistances (R_m, R_{int} and R_s, respectively). Both R_S and R_c depend on the tip contact with the sample's surface, approximated as a disk of radius a_c. As commented before, R_c has to be eliminated to independently measure the sample's thermal response. To do so, an approach developed by Spièce et al. [15] is employed, where the difference between the thermal resistances (R_{dif}) at thicknesses h and h_0 and the difference thicknesses (h_{dif}) can be evaluated with the following expression.

$$R_{dif}(h) = R_x(h) - R_x(h_0) = R_S(h) - R_S(h_0) = R_S\left(h_{dif}\right) - R_S(0) \tag{4.8}$$

This eliminates the effect of R_c as well as the related spurious effects from the local contact roughness and small surface interfacial thermal resistances caused by the BEXP or very thin oxide layers, ultimately simplifying the analysis.

The anisotropic nature of the structures can also be observed by comparing the thermal resistance versus thickness dependence in their experimental data. Specifically, the transition changes from a smooth slope for the isotropic materials to a sharper one for the most anisotropic (see Fig. 4.4c vs. d).

4.3 Analytical Modelling of the Anisotropic Thermal Transport

In this section, the experimental R_x data is treated with an analytical model to quantify the in-plane (κ_{xy}) and cross-plane (κ_z) thermal conductivities of the 2D materials for the first time [34, 35]. Also, the upper bounds of the top layers/substrates interfacial thermal insulances (r_{int}) are defined. This is done by performing a two-step data fitting. First, the data of the reference structures (RS1, RS2 and RS3) is fitted with the isotropic approach shown below to evaluate the tip/surface contact radii a_c and the dimensionless SThM tip apex correction factors α_c. These reference values are used in the second fitting step, where a modified version of the analytical model accounting for anisotropic heat transport is employed to determine the κ_{xy} and κ_z of the vdW structures. The upper bounds for the 2D materials/substrates' interfacial thermal insulances are also obtained in the second fitting, however, they do not affect the thermal conductivity values' fitting, hence limiting the fitting unknowns to two in each step. To confirm that the contact area is not physically altered after the xSThM scans, the isotropic measurements are repeated, ensuring no variation of the R_x data, hence proving the contact dimensions' stability.

The mathematical expression that models the isotropic heat transport of a top layer deposited on a substrate can be expressed in terms of spreading thermal resistance as [15, 18, 19, 36]:

$$R_S(h) = \frac{1}{\pi \kappa a_c} \int_0^{\infty} \frac{1 + \alpha_\kappa e^{\frac{-2\xi h_{eff}}{a_c}}}{1 - \alpha_\kappa e^{\frac{-2\xi h_{eff}}{a_c}}} J_1(\xi) \sin(\xi) \frac{d\xi}{\xi^2} \tag{4.9}$$

where κ is the thermal conductivity of the top layer material, J_1 is the Bessel function of the first kind and first order and ξ is the integration variable. α_κ and h_{eff} are the thermal conductivity factor and the effective thickness, respectively, and are defined as:

$$\alpha_\kappa = \left(1 - \frac{\kappa}{\kappa_s}\right) / \left(1 + \frac{\kappa}{\kappa_s}\right) \tag{4.10}$$

$$h_{eff} = h + r_{int}\kappa \tag{4.11}$$

where κ_s is the substrate's thermal conductivity. Nevertheless, this model can only be applied to fit the reference structures, as commented before, given their isotropic top layer nature (first fitting step). Hence, to fit the data of the anisotropic 2D materials (second fitting step), Eq. (4.9) is transformed to account for anisotropic heat transport by scaling the thickness into directionality-dependent values (h_m) as well as the isotropic thermal conductivity to its average anisotropic counterpart (κ_m) [18, 19].

$$(h)_{isotropic} \rightarrow (h_m)_{anisotropic} = \frac{h}{\sqrt{\kappa_z/\kappa_{xy}}} \tag{4.12}$$

$$(\kappa)_{isotropic} \rightarrow (\kappa_m)_{anisotropic} = \sqrt{\kappa_{xy}\kappa_z} \tag{4.13}$$

In order to eliminate the R_c contribution while fitting Eq. (4.9) and its modified anisotropic option with the experimental data in Eq. (4.6), the expressions have to be subtracted following Eq. (4.8). Specifically, only one curve is used in the first fitting step to obtain a_c and α_c for the top layer isotropic materials, and the same is true for the anisotropic structures, where κ_{xy} and κ_z are determined by a single data fit. These two latter parameters can be used in Eq. (2.14) to know the perpendicular anisotropic ratio ($\varphi_\perp = \kappa_{xy}\kappa_z{}^{-1}$).

This "Muzychka-Spièce" analytical approach provides insight into the variation trends of R_{dif} with respect to the xSThM parameters (Fig. 4.6). As visible in Fig. 4.6a–c, increasing the contact radius leads to smoother R_{dif} versus h_{dif} dependences and decreased R_{dif} for all the reference samples, making it possible to limit the contact radius range between 20 and 30 nm for RS1 and RS3 and between 60 and 80 nm for RS2 by comparing the experimental data with the analytical plots. Figure 4.6d–f show that the character of the R_{dif} versus h_{dif} dependences is minimally changed, mainly varying the slope angles while their "bending shape" is preserved. This is indicative of the minor effect that r_{int} has on the overall thermal properties of the studied structures, also replicated in the anisotropic layers.

The theoretical predictions shown in Fig. 4.6g–i suggest that the increase of the direction-averaged top layers' thermal conductivities (κ_m) is dictated by a monotonous drop of the thermal resistance values without a notable curve shape variation, expected as the heat transport is facilitated. When the ratio between the in-plane and cross-plane thermal conductivities of the top layer anisotropic materials ($\varphi_\perp$) increases from 1 (purely isotropic) to higher values (increased anisotropy), as shown in Fig. 4.6j–l, the R_{dif} versus h_{dif} dependences present a steeper rise (in agreement with the discussion of Fig. 4.4 in the previous section) followed by a "plateau". For instance, this is clearly visible in the plot depicting the theoretical thermal behaviour of graphite (Fig. 4.6k), where a sharper transition originates due to its high anisotropy. This "plateau" behaviour is expected due to the absence of thickness versus thermal conductance dependency for layers thicker than the tip/sample contact radius, where the substrate properties become irrelevant.

The fitting plots and their corresponding results for all the samples are shown in Fig. 4.7 and Table 4.2, respectively. The contact radius determined values are in the expected range (20 to 50 nm), except for the graphite sample, where an a_c of 62.96 nm is obtained due to previous wear of the probe. All the obtained r_{int} values of the SiO_2/silicon reference structures (Fig. 4.7a, c and e) also converge with the ones reported in the literature (~ 10^{-9} km^2 W^{-1}), enhancing the reliability of the fitting [37]. For the sensitivity coefficient α_c, the results indicate a higher rise of temperature at the apex of the probes used for the γ-InSe and perovskite samples, while the smaller value in the graphite sample represents a more homogeneously spread temperature over the probe's heater.

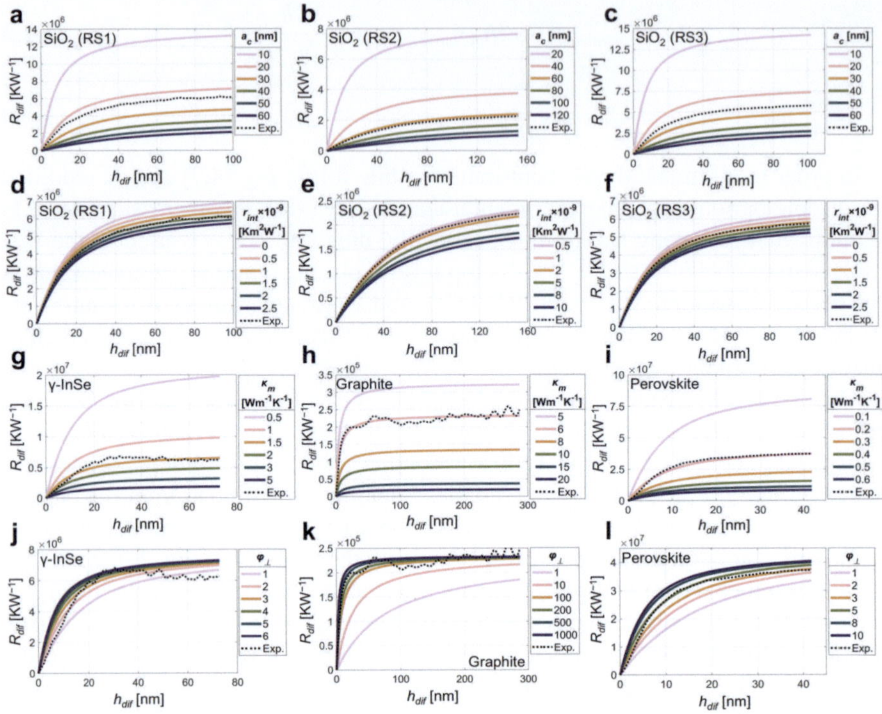

Fig. 4.6 Comparison of simulated $R_{dif}(h_{dif})$ dependences according to the analytical model (coloured lines) versus the experimental data (black dotted lines) for the different physical parameters of each of the top layer materials. **a–c** Varying a_c, **d–f** varying r_{int}, **g–i** varying κ_m and **j–l** varying $\varphi_\perp$

Once the reference values are obtained, the thermal parameters of the anisotropic materials are quantified with the second fitting (Fig. 4.7b, d and f). The fitting quality is not as accurate as in the isotropic layers, however, the minimum goodness of fit values ($R^2 > 0.94$) are sufficient to obtain confident results.

One can notice a spatial variation of the thermal signal after the transition in the experimental data, mainly in the exfoliated materials (γ-InSe and graphite), while this phenomenon is diminished or even eliminated in isotropic materials such as SiO_2 or coated two-dimensional layers (perovskite). This is thought to be the reason for the diminished fitting quality, given that exfoliated materials present staircase-like structures at their wedged surfaces, also observed in nanomechanical maps employing UFM [38, 39].

This might be indicative of additional exotic heat transport phenomena such as bulk-to-layer mode conversion or a certain degree of ballistic transport, even for thicknesses above the mean free path. Moreover, the thermal conductivities of these anisotropic materials are likely to vary when their layer thicknesses increase from thin films to bulk.

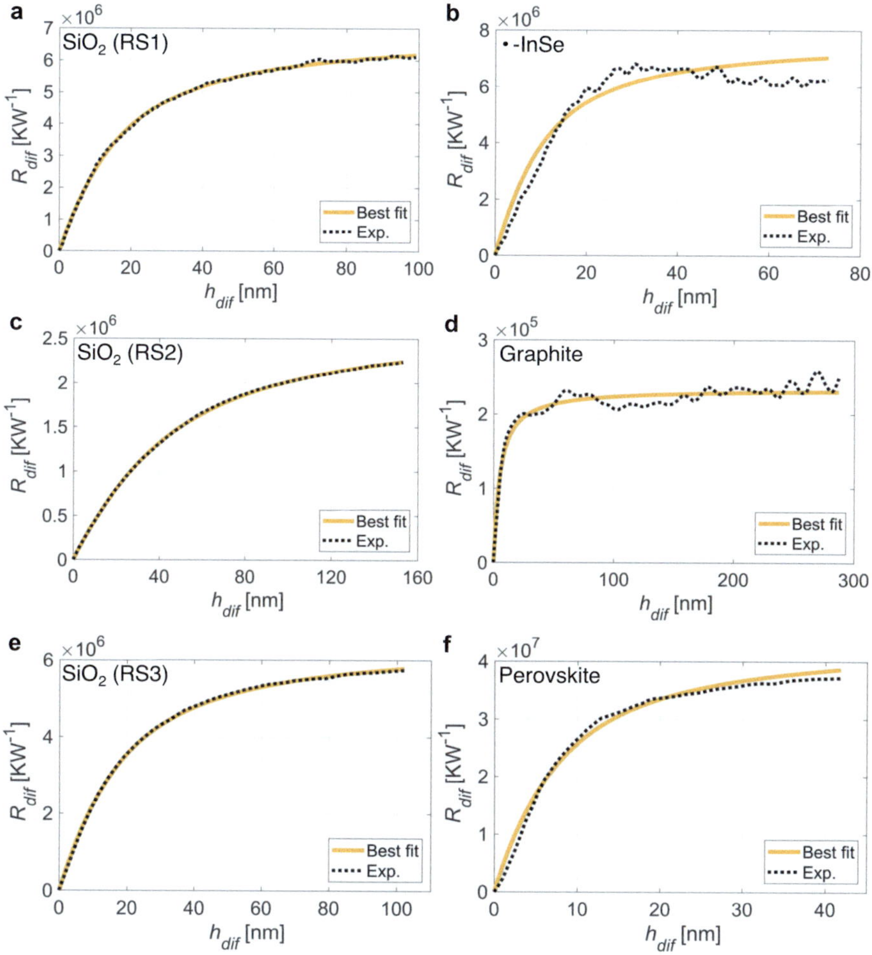

Fig. 4.7 Fitting plots of the analytical model versus the experimental data. **a**, **c** and **e** Best fits of the RS1, RS2 and RS3 top layers, respectively, and **b**, **d** and **f** best fits of the anisotropic γ-InSe, graphite and perovskite layers, respectively

To solve this accuracy problem, a separate and more detailed study would be required on similar materials beyond the measurements reported herein. Nevertheless, the interfaces between the vdW materials and their substrates do not present an observable thermal discontinuity, and the range of measured thicknesses as well as the introduced weighting factors prioritise the fitting at the initial slope and bending zone, decreasing the fitting influence of the "plateau" region. As discussed before, the initial slope-to- "plateau" transition zone is key for defining the sample's anisotropy, implying that the fitting needs to be focused there with the weighing factors.

To summarise, the xSThM measurements and the "Muzychka-Spièce" analytical approach allow to directly determine for the first time both of the absolute

Table 4.2 Fitted heat transport parameters of the studied samples

	RS1	γ-InSe/silicon	RS2	Graph./silicon	RS3	Perov./SiO_2
a_c (nm)	22.92 ± 0.28		62.96 ± 0.56		25.27 ± 0.86	
α_c	8.29 ± 0.11		1.75 ± 0.02		7.62 ± 0.19	
$r_{int} \times 10^{-11}$ (km^2 W^{-1})	148	9.60	101	1050	104	67.70
κ_{xy} (W m^{-1} K^{-1})	–	2.16 ± 0.36	–	92.29 ± 9.16	–	0.41 ± 0.04
κ_z (W m^{-1} K^{-1})	–	0.89 ± 0.09	–	0.39 ± 0.04	–	0.09 ± 0.01
$R_c \times 10^6$ (KW^{-1})	54.47 ± 0.01	45.81 ± 0.10	$6.92 \pm 2 \times 10^{-3}$	$2.54 \pm 2 \times 10^{-3}$	36.37 ± 0.10	36.01 ± 0.28

thermal conductivity components in anisotropic 2D materials with low and moderate heat conductance. This is the case for γ-InSe, where the reported values are lower compared to their literature counterparts (κ_{xy} = 10.42 W m^{-1} K^{-1} and κ_z = 1.74 W m^{-1} K^{-1}), implying enhanced thermoelectric *ZT* due to decreased heat transport ability [11, 40], as well as for the perovskite, where the obtained results match with the literature [41]. However, the obtained values of thermal conductivity for the graphite sample are one to two orders of magnitude smaller than the ones reported elsewhere (κ_{xy} = 1800 W m^{-1} K^{-1} and κ_z = 6 W m^{-1} K^{-1}) [42, 43], suggesting that the analytical model might not be entirely conclusive with highly anisotropic materials with elevated in-plane thermal conductivities or that the high thickness of the flake limits the typically enhanced thermal transport at the 2D geometry. Nevertheless, the fitted anisotropy degree is similar to the reported one.

The smaller values of the γ-InSe sample are possibly arising from the impeded phonon transport on the boundaries of the nanoscale flake. Given the phonon-dominated heat transport in vdW semiconductor materials, stiffer in-plane covalent bonds promote a greater phonon group velocity and, correspondingly, a higher in-plane thermal conductivity compared to the weaker vdW cross-plane bonds [44]. If the average thermal conductivities for each of the 2D materials is calculated via Eq. (4.13), it is possible to express the decreased overall heat transport in the γ-InSe layer (1.39 W m^{-1} K^{-1}), similar to the reported value of SiO_2. The averaged thermal conductivity of the perovskite structure results in 0.19 W m^{-1} K^{-1}, which is also in agreement with the reported values and one order of magnitude smaller than its SiO_2 substrate thermal response.

Interestingly, low interfacial thermal insulance values at the γ-InSe/silicon (r_{int} = 9.60 × 10^{-11} km^2 W^{-1}) and perovskite/SiO_2 (r_{int} = 6.77 × 10^{-10} km^2 W^{-1}) interfaces are obtained in the fitting. These small values suggest good thermal contact between the top layered materials and the substrates. However, as discussed previously and verified in the next section, the xSThM thermal conductance is generally not sensitive to interfacial thermal insulances below 10^{-9} km^2 W^{-1}, treating such value as an upper threshold of r_{int}. To confirm the feasibility of such low r_{int} values and their absence of significant effect on the xSThM results, the fittings are repeated with different interfacial thermal insulances, obtaining almost identical values for the directional thermal conductivities of both γ-InSe and perovskite materials. In the graphite sample, however, the obtained r_{int} is close below the range of values reported elsewhere [45], implying a minimally improved thermal contact between the 2D material and the silicon. The low resistance values at the three measured vdW material/substrate interfaces might arise from the plasma etching steps conducted on the substrates before the top layers' depositions, improving the adhesion of the two building blocks due to the activation of the surface and elimination of the intrinsic thin (~ 2 to 4 Å) native oxide layers.

As commented in Sect. 2.5.2, both electrons and phonons contribute to the total heat transport in a structure (κ_e and κ_L, respectively). In lightly doped γ-InSe and 2D perovskite, phonons tend to dominate the thermal conduction due to the low density of electrons (n = 10^{15} cm^{-3} in γ-InSe and n = 10^{14} cm^{-3} in perovskite) [46, 47]. Conversely, the effect of electrons in graphene's thermal conductivity is enhanced by its up to four orders of magnitude higher charge concentrations ($n = 10^{18}$ cm^{-3}) [48]. Using Wiedemann–Franz law, one can estimate the electronic component of the thermal conductivities for each of the materials as:

$$\kappa_e = \sigma LT \tag{4.14}$$

where L is the Lorenz number (2.44×10^{-8} V^2 K^{-2}). Equation (2.1) is used to determine the electrical conductivities with RT ($T = 300$ K) mobilities of $\mu = 0.10$ m^2 V^{-1} s^{-1} in γ-InSe, $\mu = 8 \times 10^{-4}$ m^2 V^{-1} s^{-1} in perovskite and $\mu = 1$ m^2 V^{-1} s^{-1} in graphene [47–49], and the results are expressed in Table 4.3. All the resulting electronic thermal conductivity values are much lower than their obtained xSThM phononic counterparts, especially in the case of the perovskite material, indicating the minimal influence of the electrons in the thermal transport of such type of structures. Also, SThM is known to be more sensitive to the phononic thermal conductivity [17], which could explain the drastic reduction of the obtained directional thermal conductivities of graphene, that has a higher electronic contribution to thermal transport.

Lastly, it is possible to calculate the interfacial thermal resistances between the tip and sample R_c (see Table 4.2). This is done by fitting Eq. (4.7) directly with the experimental data since the only unknown is R_c. As predicted, the contributions of these interfacial thermal resistances are much higher than their sample spreading resistance counterparts, hence, hindering the measurability of the intrinsic heat transport of the samples. These contact resistances can be converted to thermal insulances (r_c) to compare them to the other calculated interfaces (top layer/substrate) as:

$$r_c = R_c\left(\pi a_c^2\right) \tag{4.15}$$

where the obtained contact insulance values range from ≈ 32×10^{-9} km^2 W^{-1} at the graphite/silicon structure to ≈ 90×10^{-9} km^2 W^{-1} at the RS1.

Besides the three vdW structures exposed in this chapter, xSThM can also be applied to other types of ensembles, for instance, in antimony telluride (Sb_2Te_3/MoS_2) superlattices (discussed in Sect. 5.2) and silicon germanium (SiGe) composition gradient alloys, furtherly validating the approach [15]. Other InSe polytypes (such as ε and β) or metal halide perovskites would be interesting subjects for their xSThM studies, given that the obtained anisotropy in such structures depends on the different kinds of covalent intralayer versus. vdW interlayer bonds, resulting in a different heat transport directionality not reported elsewhere.

Table 4.3 Electron-dominated contributions to the thermal conductivity of the studied vdW structures

	γ-InSe	Graphene	Perovskite
σ ($S\ m^{-1}$)	16.02	1.60×10^5	0.01
κ_e ($W\ m^{-1}\ K^{-1}$)	1.17×10^{-4}	1.17	9.37×10^{-8}

4.4 Modelling and Validation of the Results

4.4.1 *Finite Element Analysis of Thermal Transport*

To verify the reliability of the obtained analytical results, the thermal transport in the semiconductor samples is simulated with finite element analysis (FEA) in COMSOL Multiphysics. The wedge model adapted from the real sample with a continuous thickness variation (Fig. 4.8a) is found to be less computationally-efficient in terms of simulation time, hence, a flat model of the layered materials on their substrates (Fig. 4.8b) is chosen as a more suitable FEA interface.

The difference in the heat response between the two geometries can be observed in Fig. 4.8c–e, confirming that the thermal resistances of the low angle wedge model deviate less than 0.60% with respect to the flat counterpart (highest discrepancy of 0.58% in the perovskite model). Hence, the flat model can be accurately employed to determine the parameters defining the system and compare them to the analytical results.

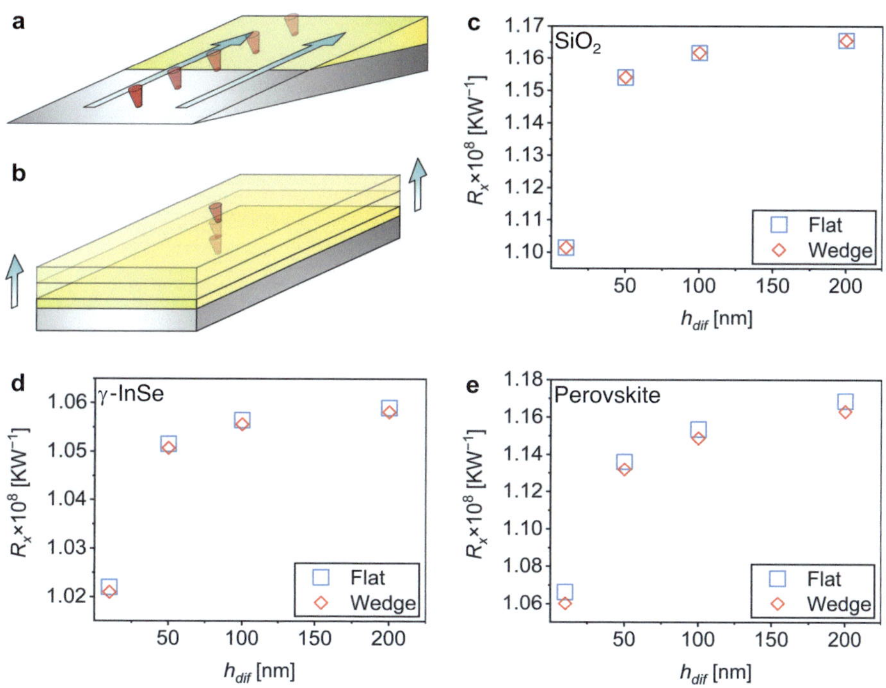

Fig. 4.8 Flat versus wedge models comparison. **a** Wedge model based on the real samples with tip movement along the surface, **b** flat model employed in the simulations with thickness variation, thermal resistance results for the wedge versus flat models' comparison at the **c** SiO_2/silicon, **d** γ-InSe/silicon and **e** perovskite/SiO_2 structures

The calculated FEA plots (similar to the theoretical ones found in Fig. 4.6) are shown in Fig. 4.9. By using linear interpolation, one can fit the experimental data with the simulated curves for the a_c parameter (Fig. 4.9a), resulting in a contact radius of 24.49 ± 1.19 nm, almost identical to the values obtained in the "Muzychka-Spièce" fitting. Additionally, Fig. 4.9b qualitatively shows the effect of r_{int} on the thermal resistance parameter, confirming that interfacial thermal insulances on the order and below 10^{-9} km^2 W^{-1} have a negligible influence on the layer's thermal transport, being consistent with the theoretical results.

The thermal simulations of the anisotropic semiconductor structures are shown in Fig. 4.9c–f. For the curves varying the averaged anisotropic thermal conductivities, the same linear interpolation can be applied to the experimental data to find values of $\kappa_m = 1.17 \pm 0.48$ W m^{-1} K^{-1} at the γ-InSe layer and $\kappa_m = 0.13 \pm 0.07$ W m^{-1} K^{-1} at the perovskite layer, also close to the analytical results. On the other hand, the

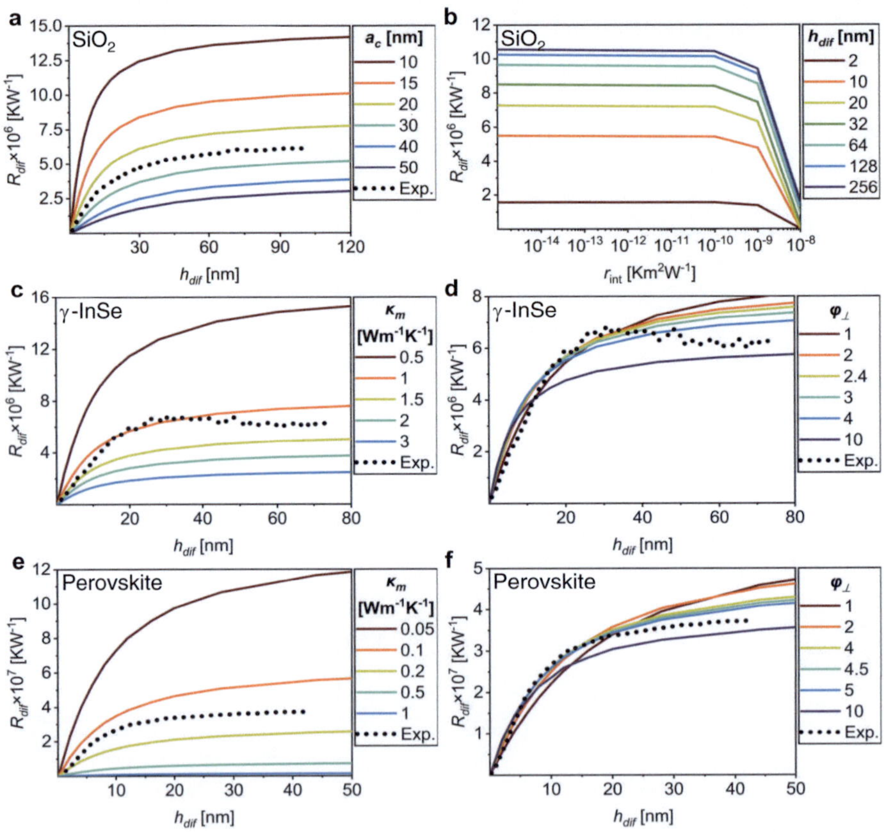

Fig. 4.9 FEA simulation plots (coloured lines) versus the experimental data (black dotted lines) for the different simulated physical parameters. **a** Varying a_c and **b** varying r_{int} in the SiO_2/silicon layers, **c**, **e** varying κ_m and **d**, **f** varying $\varphi_\perp$ in the γ-InSe/silicon and perovskite/SiO_2 layers, respectively

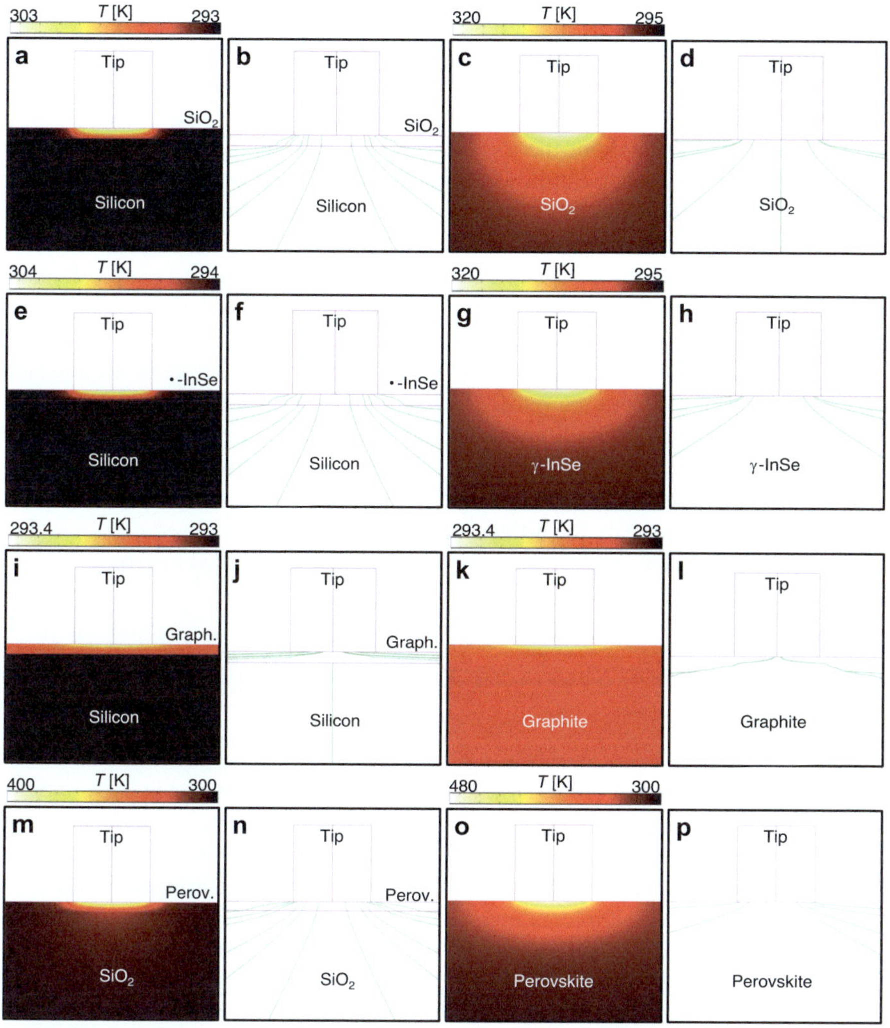

Fig. 4.10 FEA thermal rendered plots. For the SiO_2/silicon structure: **a**, **c** temperature and **b**, **d** heat flow plots for the thin and thick top layers, respectively. For the γ-InSe/silicon structure: **e**, **g** temperature and **f**, **h** heat flow plots for the thin and thick top layers, respectively. For the graphene/silicon structure: **i**, **k** temperature and **j**, **l** heat flow plots for the thin and thick top layers, respectively. For the perovskite/SiO_2 structure: **m**, **o** temperature and **n**, **p** heat flow plots for the thin and thick top layers, respectively

simulated plots showing the variation of the anisotropic ratio (Fig. 4.9d and f) have a less clear comparison to the experimental data, nevertheless, iterative fitting can be applied by fixing the averaged thermal conductivities formerly obtained, providing ratios of 2.40 and 4.50 for the γ-InSe and perovskite layers, respectively (analogous as well to the previous section values).

COMSOL also allowed to render qualitative maps of the heat flow and temperature distribution inside the samples. In this case, all the materials of interest are compared in Fig. 4.10 for both thin (5 nm) and thick (200 nm) layers.

The proximity of the substrate to the heat source (SThM tip) has a vital role in the heat spreading throughout the top thin layers, behaving as a heat sink rapidly dissipating any heat concentration when the substrate is a good thermal conductor (silicon) and restricting the phonon flow accumulating a temperature gradient in ensembles containing low thermal conductivity substrates (SiO_2). Such behaviour is not as evident in the thick layer plots, where the substrate is away from the heat source and the true nature of the anisotropic heat flow is shown in the top layer materials.

The increased anisotropy can be clearly observed in both the temperature and heat flow streamline plots, from the purely isotropic SiO_2 indicating a multidirectional thermal behaviour, to the highly anisotropic graphene where the streamlines are mostly in-plane oriented. The heat flux in γ-InSe and perovskite is significantly similar due to their equivalent averaged thermal conductivities, also being analogous to the one in SiO_2. However, the difference in anisotropy can be observed as a higher in-plane versus cross-plane heat transport in perovskite compared to γ-InSe.

4.4.2 *Evaluation of Systematic Error with an Isotropic Structure*

Given the unique nature of the xSThM technique, the results have been corroborated with an isotropic sample of known thermal properties. In this case, a polymeric SU-8 layer ($\kappa = 0.20$ W m^{-1} K^{-1}) [50] is fabricated and analysed replicating the process described in the previous sections to evaluate a possible systematic error. Figure 4.11 depicts the xSThM topographical and thermal response as well as the fitting plots of such structure. The xSThM thermal map and profile show a qualitative decrease of the thermal transport in SU-8 compared to the substrate layers, indicated by a higher thermal signal (Fig. 4.11b and d). The system parameters are fitted following the two-step "Muzychka-Spièce" process, resulting in an averaged top layer thermal conductivity of $\kappa_m = 0.22 \pm 0.01$ W m^{-1} K^{-1} and an anisotropic ratio of $\varphi_\perp = 0.95$, proving the material's isotropic nature as well as the low xSThM error ($< 7.50\%$).

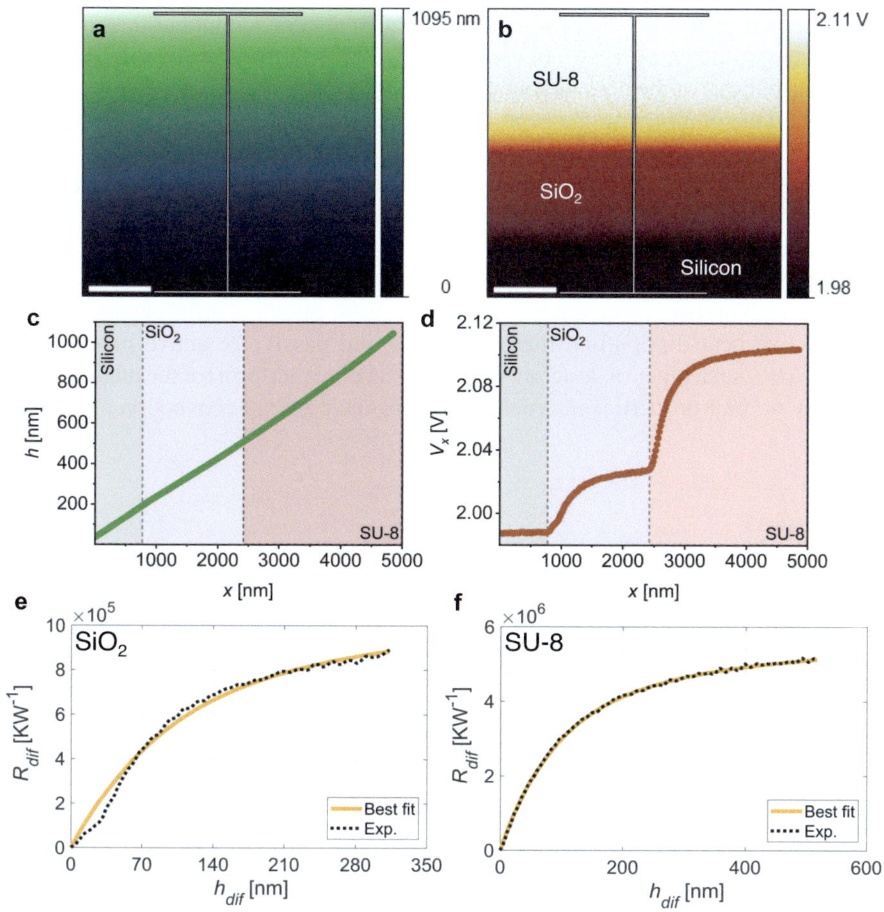

Fig. 4.11 xSThM results for the isotropic SU-8 sample. **a** Topography and **b** thermal signal maps of the BEXP structure (scale bars are 1 μm), **c** topography and **d** thermal signal profiles extracted from **a** and **b**, respectively, and **e** SiO_2 and **f** SU-8 layers fitting plots

4.5 Summary

The independent measurements of the heat transport parameters of different 2D materials have been addressed in this chapter. A novel methodology of cross-sectional scanning thermal microscopy has given access to the precise quantification of the in-plane and cross-plane thermal conductivities of such structures as well as the evaluation of the upper bounds of their interfacial thermal insulances with the substrates. This approach eliminates the need for the deposition of additional layers present in other types of techniques, facilitating the measurements. The performed BEXP fabrication nano-polishes the samples at a near-glancing angle to access the internal layers with minimal damage. This is followed by a nanoscale resolution SThM mapping of the thermal resistance of the scanned structures at varying thickness. An analytical model (validated by FEA simulations) is employed to determine the independent contributions of the characteristically directional thermal conductivities in such vdW materials, eliminating the need to determine the unknown tip/sample thermal resistance.

The anomalously low obtained thermal conductivity values prove the applicability of the studied semiconductor materials (γ-InSe and perovskite) for thermoelectric or optoelectronic technologies. Overall, the true nanoscale lateral precision of xSThM is crucial for the understanding of the thermal characteristics of ultrathin anisotropic vdW materials and their heterostructures as well as for the application of such systems in fields like heat dissipation, thermoelectrics and passive or active nanothermal devices. Also, such types of structures provide a tunable platform for the manipulation and study of their properties in order to enhance their energy conversion efficiency.

References

1. Lin, Z., McNamara, A., Liu, Y., Moon, K.-S., & Wong, C.-P. (2014). Exfoliated hexagonal boron nitride-based polymer nanocomposite with enhanced thermal conductivity for electronic encapsulation. *Composites Science and Technology, 90*, 123–128.
2. Zhang, Z., Hu, S., Chen, J., & Li, B. (2017). Hexagonal boron nitride: A promising substrate for graphene with high heat dissipation. *Nanotechnology, 28*, 225704.
3. Xu, Y., Li, Z. Y., & Duan, W. H. (2014). Thermal and thermoelectric properties of graphene. *Small (Weinheim an der Bergstrasse, Germany), 10*, 2182–2199.
4. Amollo, T. A., Mola, G. T., Kirui, M. S. K., & Nyamori, V. O. (2018). Graphene for thermoelectric applications: Prospects and challenges. *Critical Reviews in Solid State and Materials Sciences, 43*, 133–157.
5. Pallecchi, I., Manca, N., Patil, B., Pellegrino, L., & Marre, D. (2020). Review on thermoelectric properties of transition metal dichalcogenides. *Nano Futures, 4*, 032008.
6. Zhou, S., Tao, X., & Gu, Y. (2016). Thickness-dependent thermal conductivity of suspended two-dimensional single-crystal In_2Se_3 layers grown by chemical vapor deposition. *Journal of Physical Chemistry C, 120*, 4753–4758.
7. Wu, J., Chen, Y. B., Wu, J. Q., & Hippalgaonkar, K. (2018). Perspectives on thermoelectricity in layered and 2D materials. *Advanced Electronic Materials, 4*, 1800248.
8. Mak, K. F., Lee, C., Hone, J., Shan, J., & Heinz, T. F. (2010). Atomically thin MoS_2: A new direct-gap semiconductor. *Physical Review Letters, 105*, 136805.
9. Shi, H., Wang, D., Xiao, Y., & Zhao, L.-D. (2021). Dynamic carrier transports and low thermal conductivity in n-type layered InSe thermoelectrics. *Aggregate, 2*, e92.
10. Yang, S. J., et al. (2021). Enhancing thermoelectric power factor of 2D organometal halide perovskites by suppressing 2D/3D phase separation. *Advanced Materials, 33*, 2102797.
11. Buckley, D., et al. (2021). Anomalous low thermal conductivity of atomically thin InSe probed by scanning thermal microscopy. *Advanced Functional Materials, 31*, 2008967.
12. Bandurin, D. A., et al. (2017). High electron mobility, quantum Hall effect and anomalous optical response in atomically thin InSe. *Nature Nanotechnology, 12*, 223–227.
13. Rasel, M. A. J., et al. (2020). Chain-length dependence of thermal conductivity in 2D alkylammonium lead iodide single crystals. *ACS Applied Materials & Interfaces, 12*, 53705–53711.
14. Tovee, P., Pumarol, M., Zeze, D., Kjoller, K., & Kolosov, O. (2012). Nanoscale spatial resolution probes for scanning thermal microscopy of solid state materials. *Journal of Applied Physics, 112*, 114317.

15. Spièce, J., et al. (2021). Quantifying thermal transport in buried semiconductor nanostructures via cross-sectional scanning thermal microscopy. *Nanoscale, 13*, 10829–10836.
16. Spiece, J., et al. (2018). Improving accuracy of nanothermal measurements via spatially distributed scanning thermal microscope probes. *Journal of Applied Physics, 124*, 015101.
17. Evangeli, C., et al. (2019). Nanoscale thermal transport in 2D nanostructures from cryogenic to room temperature. *Advanced Electronic Materials, 5*, 1900331.
18. Muzychka, Y. S., Yovanovich, M. M., & Culham, J. R. (2004). Thermal spreading resistance in compound and orthotropic systems. *Journal of Thermophysics and Heat Transfer, 18*, 45–51.
19. Muzychka, Y. S. (2014). Spreading resistance in compound orthotropic flux tubes and channels with interfacial resistance. *Journal of Thermophysics and Heat Transfer, 28*, 313–319.
20. Gomès, S., Assy, A., & Chapuis, P.-O. (2015). Scanning thermal microscopy: A review. *Physica Status Solidi A, 212*, 477–494.
21. Timofeeva, M., et al. (2016). Scanning thermal microscopy with heat conductive nanowire probes. *Ultramicroscopy, 162*, 42–51.
22. Zhu, W., Zheng, G., Cao, S., & He, H. (2018). Thermal conductivity of amorphous SiO_2 thin film: A molecular dynamics study. *Science and Reports, 8*, 10537.
23. Brotzen, F. R., Loos, P. J., & Brady, D. P. (1992). Thermal conductivity of thin SiO_2 films. *Thin Solid Films, 207*, 197–201.
24. Shanks, H. R., Maycock, P. D., Sidles, P. H., & Danielson, G. C. (1963). Thermal conductivity of silicon from 300 to 1400°K. *Physical Review, 130*, 1743–1748.
25. Li, X., et al. (2017). Thermal conduction across a boron nitride and SiO_2 interface. *Journal of Physics D: Applied Physics, 50*, 104002.
26. Bosse, J. L., Grishin, I., Huey, B. D., & Kolosov, O. V. (2014). Nanomechanical morphology of amorphous, transition, and crystalline domains in phase change memory thin films. *Applied Surface Science, 314*, 151–157.
27. Bosse, J. L., et al. (2014). Nanothermal characterization of amorphous and crystalline phases in chalcogenide thin films with scanning thermal microscopy. *Journal of Applied Physics, 116*, 134904.
28. Tovee, P. D., et al. (2014). Nanoscale resolution scanning thermal microscopy using carbon nanotube tipped thermal probes. *Physical Chemistry Chemical Physics: PCCP, 16*, 1174–1181.
29. Song, J., Zhou, Y., Padture, N. P., & Huey, B. D. (2020). Anomalous 3D nanoscale photoconduction in hybrid perovskite semiconductors revealed by tomographic atomic force microscopy. *Nature Communications, 11*, 3308.
30. Zhang, W., et al. (2023). Tunable electrical field-induced metal-insulator phase separation in $LiCoO_2$ synaptic transistor operating in post-percolation region. *Nano Energy, 108*, 108199.
31. Handa, T., Yamada, T., Nagai, M., & Kanemitsu, Y. (2020). Phonon, thermal, and thermo-optical properties of halide perovskites. *Physical Chemistry Chemical Physics: PCCP, 22*, 26069–26087.
32. Mafakheri, M., & Dodaran, A. A. S. (2021). Graphene phonons lifetime and mean free path using ab initio molecular dynamics and spectral energy density analysis. *Physica Scripta, 96*, 065702.
33. Li, H. L., & Cao, B. Y. (2019). Radial ballistic-diffusive heat conduction in nanoscale. *Nanoscale and Microscale Thermophysical Engineering, 23*, 10–24.
34. Gonzalez-Munoz, S., et al. (2023). Direct measurements of anisotropic thermal transport in γ-InSe nanolayers via cross-sectional scanning thermal microscopy. *Advanced Materials Interfaces, 10*, 2300081.
35. Maiti, A., Agarwal, K., Gonzalez-Munoz, S., & Kolosov, O. V. (2023). Quantifying anisotropic thermal transport in two-dimensional perovskite (PEA_2PbI_4) through cross-sectional scanning thermal microscopy. *Physical Review Materials, 7*, 023801.
36. Yovanovich, M. M., Culham, J. R., & Teertstra, P. (1998). Analytical modeling of spreading resistance in flux tubes, half spaces, and compound disks. *EEE Transactions on Components, Packaging, and Manufacturing Technology, Part A, 21*, 168–176.
37. Chen, J., Zhang, G., & Li, B. (2012). Thermal contact resistance across nanoscale silicon dioxide and silicon interface. *Journal of Applied Physics, 112*, 064319.

38. Tapia-Ruiz, N., et al. (2021). 2021 roadmap for sodium-ion batteries. *Journal of Physics: Energy, 3*, 031503.
39. Bosse, J. L., Tovee, P. D., Huey, B. D., & Kolosov, O. V. (2014). Physical mechanisms of megahertz vibrations and nonlinear detection in ultrasonic force and related microscopies. *Journal of Applied Physics, 115*, 144304.
40. Atakishiev, S. M., Abdinov, D. S., & Akhundov, G. A. (1968). Heat conductivity of indium selenide. *Physica Status Solidi B, 28*, K47–K50.
41. Li, C., et al. (2021). Remarkably weak anisotropy in thermal conductivity of two-dimensional hybrid perovskite butylammonium lead iodide crystals. *Nano Letters, 21*, 3708–3714.
42. Lee, J. U., Yoon, D., Kim, H., Lee, S. W., & Cheong, H. (2011). Thermal conductivity of suspended pristine graphene measured by Raman spectroscopy. *Physical Review B, 83*, 081419.
43. Pop, E., Varshney, V., & Roy, A. K. (2012). Thermal properties of graphene: Fundamentals and applications. *MRS Bulletin, 37*, 1273–1281.
44. Vialla, F., & Del Fatti, N. (2020). Time-domain investigations of coherent phonons in van der Waals thin films. *Nanomaterials, 10*, 2543.
45. Zhang, J., Wang, Y., & Wang, X. (2013). Rough contact is not always bad for interfacial energy coupling. *Nanoscale, 5*, 11598–11603.
46. Milot, R. L., et al. (2016). Charge-carrier dynamics in 2D hybrid metal–halide perovskites. *Nano Letters, 16*, 7001–7007.
47. Zaslonkin, A. V., Kaminskii, V. M., Kovalyuk, Z. D., Mintyanskii, I. V., & Tovarnitskii, M. V. (2006). Electrical properties of hydrogenated InSe crystals. *Inorganic Materials, 42*, 1308–1310.
48. Novoselov, K. S., et al. (2005). Two-dimensional gas of massless Dirac fermions in graphene. *Nature, 438*, 197–200.
49. Kober-Czerny, M., et al. (2022). Excellent long-range charge-carrier mobility in 2D perovskites. *Advanced Functional Materials, 32*, 2203064.
50. Seok Hwan, O., Kwang-Cheol, L., Jaechul, C., Moohwan, K., & Seung, S. L. (2001). Micro heat flux sensor using copper electroplating in SU-8 microstructures. *Journal of Micromechanics and Microengineering, 11*, 221.

Chapter 5
Thermoelectricity in 2D–3D Nanostructures

Abstract This chapter builds upon the thermal transport insights gained in the previous chapter and integrates them with electronic characterisation to explore thermoelectric performance in hybrid 2D–3D nanostructures. The main objective is to demonstrate how interfacial engineering can be used to simultaneously optimise charge and heat transport, thereby enhancing thermoelectric efficiency at room temperature. The first system investigated is a molybdenum disulphide/antimony telluride superlattice, where the structural composition is examined in detail, followed by measurements of its electronic transport properties and anisotropic thermal conductivities. This layered structure leverages the contrast between the electronic and phononic behaviours of its constituents to tailor the thermoelectric response. The second system is a silver/antimony telluride nanocomposite, analysed similarly through its electronic and thermal transport characteristics. In both cases, the combined analysis of the electrical and thermal conductivities as well as their interfacial interactions allows for the evaluation of the materials' potential for energy harvesting applications. By correlating structural design with transport measurements, this chapter demonstrates the viability of low-dimensional heterostructures as platforms for thermoelectric optimisation, paving the way for more efficient solid-state energy conversion technologies.

5.1 Introduction

Determining the thermal properties of layered materials is only the first step towards validating their prospectively enhanced thermoelectric (TE) behaviour. It must be noted that a suitable TE device should own a combination of diminished heat transport and augmented electronic capabilities, a rather rare phenomenon due to the direct coupling of the representative parameters. Hence, the TE efficiency, denoted as ZT, tends to be low in the majority of the structures. To engineer a proper TE energy reutilisation, multiple approaches have been proposed in the literature.

Amongst them, 2D–3D superlattices and 2D-based mixed-phase nanocomposites represent viable platforms to accomplish such task [1]. The characteristic low

S. Gonzalez-Munoz, *Interplay of Nanoscale Physical Properties in 2D Materials and Heterostructures*, Springer Theses, https://doi.org/10.1007/978-3-032-14271-9_5

dimension and anisotropy of such systems are the key aspects for the decoupling of their quantum electronic and thermal properties, ultimately improving their TE performance [2, 3]. Specifically, the unique size-dependent electronic structure in 2D materials has an impact on the Seebeck coefficient and charge transport, while the direction-related scattering boundaries restrict the thermal propagation at a quantum scale. Moreover, these 2D structures allow high integration and power densities in microelectronic systems. However, many of the structures reported having a high *ZT* can only reach such efficiency at elevated temperatures (> 500 K), compromising the overall system functionality. Also, other power generation methods with higher efficiencies are more appropriate to use in such high temperature ranges. Hence, research is being focused on TE structures capable of performing energy conversion at moderate temperature ranges (300–500 K) while maintaining a proper *ZT*, with the highest reported RT *ZT* of ≈ 2.40 in a Bi_2Te_3/Sb_2Te_3 superlattice [4]. The excellent TE properties of these two compositions can be prospectively optimised by multilayer stacking with MoS_2, given the favourable predictions of its DOS and electronic behaviour, capable of enhancing the Seebeck effect [4, 5]. Silver doping can also aid in the reduction of the lattice thermal conductivity given the modified local interatomic bonding, yielding a strong phonon scattering [6].

In this chapter, this enhancement of *ZT* is analysed in both Sb_2Te_3/MoS_2 superlattices as well as $Sb_2Te_3/AgSbTe_2$ hybrid nanocomposites, with a deep focus on their nanoscale electronic characteristics and their anisotropic thermal transport. Sb_2Te_3 is a widely studied topological insulator with favourable TE features [1]. The superlattice stacking and nanocomposite doping lead to synergistic band modulation affecting the electronic measurements. The xSThM analytical approach described in the previous chapter is also employed here to extract the anisotropic heat features of the TE structures for the first time, improving the accuracy and reliability of the nanoscale results compared to other methods limited to bulk materials and unable to deconvolute the directionality in the phononic transport. Such thermal measurements of the nanoscale layers combined with their in-plane electronic characterisation allow to confirm increased values of power factor and reduced thermal conductivities indicating an enhanced *ZT* efficiency at the lower side of the moderate temperature range. Therefore, the results achieved in this chapter set the foundation of a versatile methodology capable of fully exploiting the TE properties of 2D–3D systems.

The fabrication, composition and electronic measurements of the samples explained in this chapter were performed in collaboration with Prof. Bodh Raj Mehta, Mujeeb Ahmad and Abhishek Ghosh from the Indian Institute of Technology Delhi. The BEXP and thermal analysis were performed by Dr. Khushboo Agarwal and myself in the Lancaster University Physics Department. The results exposed in this chapter have been published in Ahmad, M. et al. *Adv. Funct. Mater.* **32** (2022) and Ghosh, A. et al. (2002) (see List of publications).

5.2 Molybdenum Disulphide/Antimony Telluride Superlattice

5.2.1 Structural Composition

Two multilayer samples are fabricated via multitarget radiofrequency sputtering of 99.90% pure Sb_2Te_3 and MoS_2 sequential thin films on SiO_2/silicon substrates, forming superlattices. An additional pristine Sb_2Te_3 sample is fabricated following a similar single-step sputtering. Prior to the deposition, the substrates are cleaned following the standard Radio Corporation of America (RCA) process. During the sputtering, the P_{RF} is kept at 20 and 50 W for the Sb_2Te_3 and MoS_2 layers, respectively, while the argon flow, the substrate temperature and the chamber pressure are 20 SCCM, 150 °C and 7.33×10^{-3} mbar, respectively. A 10 min presputtering is conducted to eliminate any remaining contamination before the deposition starts. Such deposition is held at rates of 10 and 2 nm min^{-1} for the Sb_2Te_3 and MoS_2, respectively. In the two multilayer samples, the thickness of Sb_2Te_3 is fixed at 20 nm, while the MoS_2 layers vary from 5 nm in one sample (Sb_2Te_3/M5) to 10 nm in the other (Sb_2Te_3/M10). These deposition sequences are repeated until each sample forms a thin film consisting of 10 layers of MoS_2 and 11 layers of Sb_2Te_3 over the substrate, adding up to a total thickness of 270 nm in the Sb_2Te_3/M5 sample and 320 nm in the Sb_2Te_3/M10 sample. Before the thermal analysis, the samples are also cut via BEXP in order to access the buried atomically flat wedge-like surfaces.

After the Raman spectroscopy measurements discussed in Sect. A.2.5, structural characterisation is conducted on the top surface of the thin films, as observed in Fig. 5.1a–c. In these images taken with a dual beam field emission SEM (FEI Quanta 3D FEG), the morphology variation can be distinguished between the three samples. The pristine Sb_2Te_3 sample contains the largest nanocrystalline grains (200–300 nm), while the introduction of MoS_2 modifies the size of the top surface elements and smoothens the plane. The average grain size of the multilayer samples is measured to be 100–120 and 40–60 nm for the Sb_2Te_3/M5 and Sb_2Te_3/M10 thin films, respectively. Such reduction in the grain dimensions originates from the lower thickness of Sb_2Te_3 in the multilayer samples and the lattice mismatch with MoS_2, which is expected to affect the Seebeck coefficient and thermal conductivity [7].

Additionally, to verify the thicknesses of the stacked layers, sectional high resolution TEM (FEI Tecnai G^2 20 S-TWIN) is conducted on the superlattices (Fig. 5.1d), distinctively confirming the stated values of 20, 10 and 5 nm for the Sb_2Te_3, M10 and M5 layers, respectively. The fringed pattern in the Sb_2Te_3 layers indicates its crystalline phase with an interplanar distance of 1 nm, whereas the absence of a similar pattern in MoS_2 denotes that the composition of such layers is comprised of isolated domains with relatively poor crystallinity.

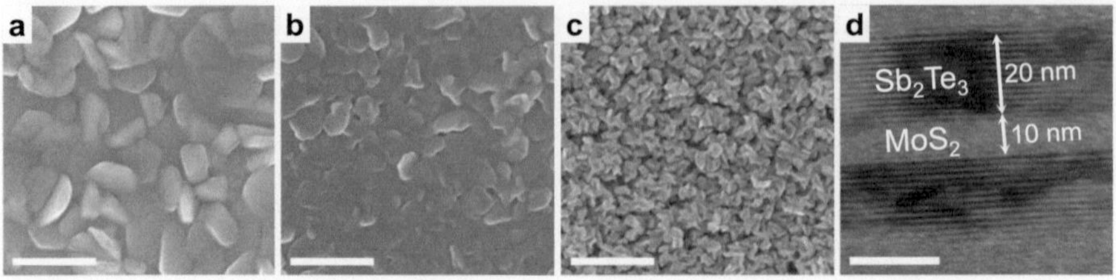

Fig. 5.1 Top view field emission SEM micrographs of the **a** Sb_2Te_3, **b** Sb_2Te_3/M5 and **c** Sb_2Te_3/M10 thin films (scale bars are 200 nm) and **d** sectional high resolution TEM image of the Sb_2Te_3/M10 sample (scale bar is 20 nm)

5.2.2 Electronic Transport Characteristics

In-plane Hall measurements (Quantum Design PPMS®) and experimental TE characterisation (Linseis LSR-3) are performed to elucidate the charge transport nature of the three samples, with the quantitative results shown in Table 5.1. An inverse dependence is observed between the carrier concentration and the Hall mobility as the MoS_2 layers increase in thickness, attributing such enhancement in mobility to the (001) preferential charge transport route in MoS_2 and the low thickness of the Sb_2Te_3 topological insulator layers [8]. Moreover, S is directly proportional to the carriers' mobility and inversely dependent on the concentration [9], while, as seen in Eq. (2.1), σ is directly proportional to both of these parameters. Thus, the engineering of the power factor ($S^2\sigma$) requires an optimised carrier concentration with increased mobility.

The in-plane experimental results are qualitatively plotted in a range of moderate temperatures (Fig. 5.2a–c). For the Sb_2Te_3 sample, the typical semiconductor increase of σ versus T is observed until ~ 400 K, then, the relation saturates for the rest of the temperature range. However, the variation of σ versus T is observed to be negligible in the multilayer samples. Also, σ decreases as more MoS_2 is introduced in the system, owing to the decrease in the carrier concentration of the higher MoS_2 thickness layers and the grain size reduction of the thin film [10].

A p-type semiconductor behaviour of the samples can be derived from the positive Seebeck coefficients shown in Fig. 5.2b, where the enhanced S range can also be observed comparing the multilayer samples versus the pristine one. Interestingly, as

Table 5.1 In-plane electronic transport parameters of the studied samples at 316 K

	Sb_2Te_3	Sb_2Te_3/M5	Sb_2Te_3/M10
$n \times 10^{19}$ (cm^{-3})	13.50 ± 0.70	4.80 ± 0.10	1.90 ± 0.08
μ ($cm^2\ V^{-1}\ s^{-1}$)	27 ± 1	46 ± 2	58 ± 3
σ ($S\ cm^{-1}$)	597 ± 29	356 ± 17	176 ± 8
S ($\mu V\ K^{-1}$)	188 ± 9	286 ± 14	530 ± 26
$S^2\sigma$ ($m\ W\ m^{-1}\ K^{-2}$)	2.11 ± 0.17	2.91 ± 0.24	4.94 ± 0.43

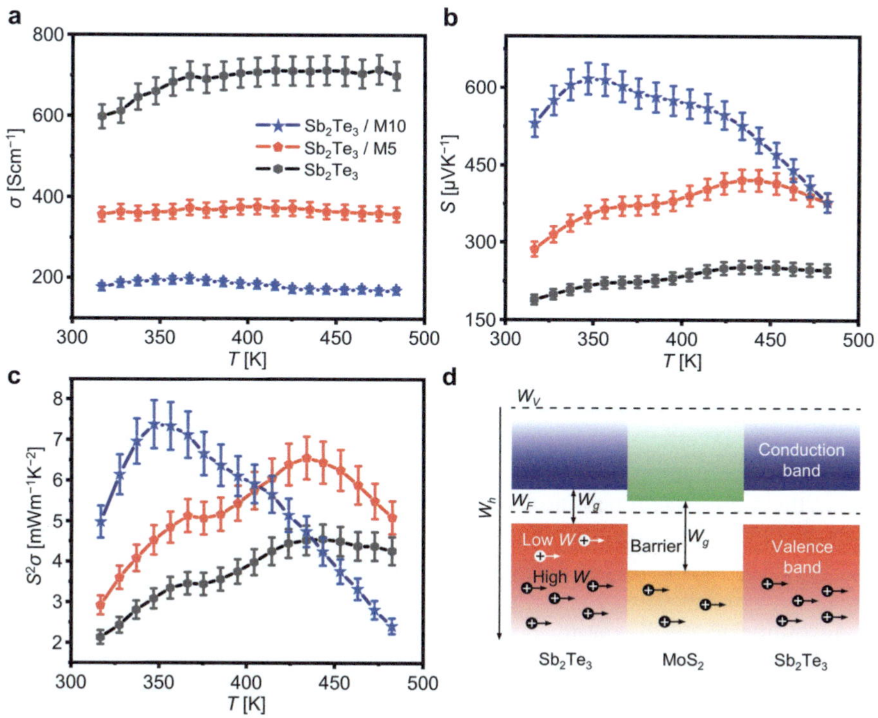

Fig. 5.2 In-plane electronic characteristics of the studied samples. **a** Electrical conductivities, **b** Seebeck coefficients, **c** power factors against temperature and **d** band diagram showing the charge transport. Notice that W_h, W_V, W_F and W_g represent the hole, vacuum, Fermi and bandgap energies, respectively

temperature increases, the corresponding S values of Sb_2Te_3 reach a saturation at around 450 K, but, in the M5 multilayer sample, S decreases proportionally after 433 K. Similarly, the M10 sample also shows a decrease, however it happens sooner (347 K) and in a steeper trend. Such reduction of the Seebeck coefficient at higher temperatures could arise from the increased contribution of minority charge carriers such as electrons [11].

With the known p-type behaviour of Sb_2Te_3 and n-type properties of MoS_2, one can assume that most of the charge contribution in the Seebeck coefficient will originate from holes in Sb_2Te_3 and electrons in MoS_2 at high temperatures, creating bipolar thermo-diffusion and reducing S.

Figure 5.2d shows the potential barrier produced at the interfaces of the superlattices due to the different work function of the involved materials. A charge excess in MoS_2 is created given its low thickness, hindering the transport of the positively charged low energy holes scattered in the heterointerfaces, a phenomenon known as carrier filtering effect (see Sect. 2.5.2) [12]. Such behaviour potentiates the contribution of high energy carriers only, greatly enhancing the Seebeck coefficients and decreasing the electrical conductivity due to the multiple interfaces present in the samples.

The resulting power factor values are enhanced in the multilayer samples given the increased S ranges and carrier mobilities, with an elevated $S^2\sigma$ value of 4.94 ± 0.43 m W m^{-1} K^{-2} observed at 316 K in the Sb_2Te_3/M10 sample.

5.2.3 Heat Transport Measurements

Given the geometrical nature of the studied samples, the xSThM heat transport analysis applied in the previous chapter can also be used here to precisely determine the anisotropic thermal properties of the superlattice structures as well as calculate their final in-plane TE efficiency. After the BEXP provides in situ access to a continuous inner surface of varying thickness of the top layers' and substrates' materials as well as their interfaces, one can scan the topographical and thermal response of such layers and extract the corresponding profiles, as shown in Fig. 5.3. The experimental data is directly exported from the profiles to Eq. (4.6) in order to convert it to thermal resistance.

According to the literature, Sb_2Te_3 alone has a higher directionally averaged thermal conductivity than its SiO_2 substrate ($\kappa_m = 3.08$ W m^{-1} K^{-1}), shown as a decrease of the thermal signal as the layer thickness is increased in Fig. 5.3c instead of the typical increase shown in all the other structures (after the shifting of the probe position to ensure a full contact of the tip with the top layer structure) [13]. This results in thermal resistance values widely scattered and not suitable for the fitting model used here, which requires a cohesive set of data increasing after the interface. Hence, the thermal values for the pristine sample are obtained via a conventional HV four-probe 3ω-method with an additional Wheatstone bridge setup to maximise the resistive heating, given that the uniformity and high thickness of the top layer (> 200 nm) allows to extract reliable results. However, the multilayer samples, composed of a stack of Sb_2Te_3 and MoS_2 thin films, are much less structurally uniform, making their 3ω-method analysis not feasible. Thus, only the resulting anisotropic experimental data of the superlattice structures can be compared to the theoretical model known as the "Muzychka-Spièce" approach, as shown in Fig. 5.4. From this figure, it is possible to do a preliminary estimation of the average thermal conductivities, positioning κ_m between 0.50 and 1 W m^{-1} K^{-1} for both superlattice samples, indicating their lower thermal conductivities compared to the SiO_2 substrates and, together with their increasing trend, validating the application of such fitting model for the evaluation of the anisotropic thermal transport.

Given that the phononic mean free path in the top layer structures is smaller than the dimensions of the tip/sample contact and the thicknesses of such layers, the diffusion approximation can be used in the analytical fitting of the thermal data, with the results shown in Table 5.2 [13]. As commented in the previous chapter, such process starts with an isotropic fitting employing the data from the SiO_2/silicon layers to obtain the contact radius (a_c) and the probe temperature gradient correction factor

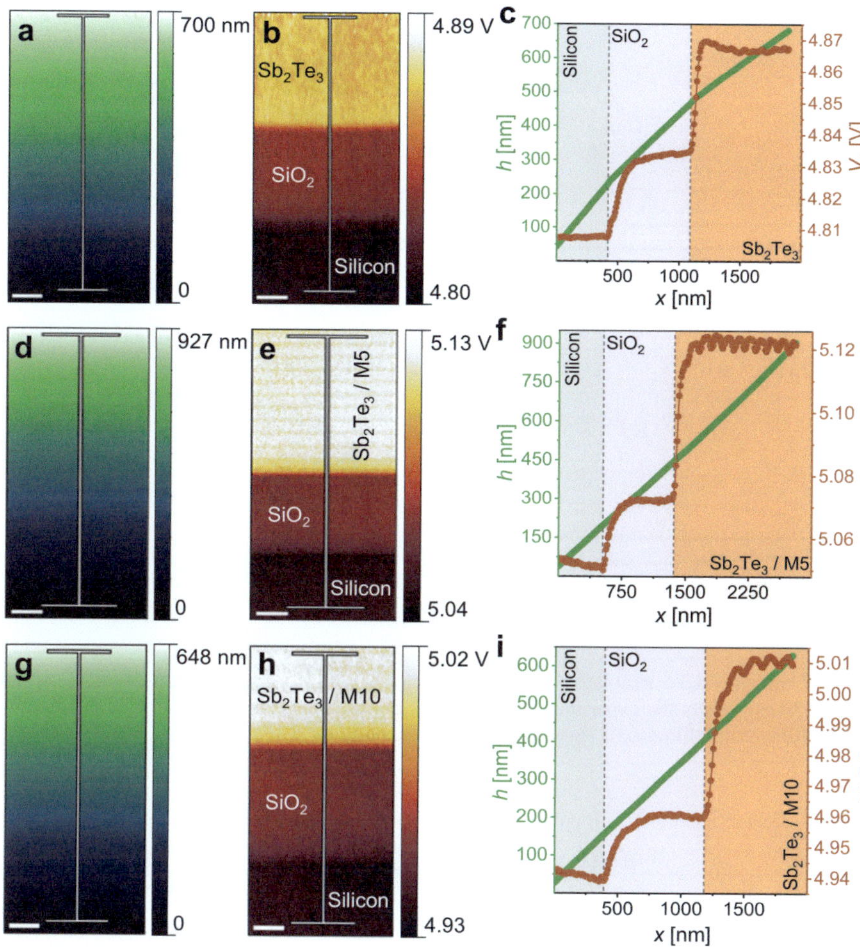

Fig. 5.3 xSThM maps and profiles of the analysed samples. **a**, **d**, **g** Topography maps, **b**, **e**, **h** thermal signal maps (scale bars are 200 nm) and **c**, **f**, **i** corresponding profiles of the Sb_2Te_3, Sb_2Te_3/M5 and Sb_2Te_3/M10 samples, respectively

(α_c). In this case, the thermal conductivities for the SiO_2 ($\kappa = 1.40$ W m^{-1} K^{-1}) and silicon ($\kappa_s = 130$ W m^{-1} K^{-1}) layers as well as their interfacial thermal insulance r_{int} ($r_{int} = 10^{-9}$ km^2 W^{-1}) are fixed using the values extracted from the literature [14–16]. The interfacial thermal insulance can be assumed to be constant given that the substrates are the same as the ones described in the previous chapter, which resulted in r_{int} values in agreement with the literature.

Once the initial fitting is performed with Eqs. (4.8) and (4.9), one can obtain the anisotropic thermal conductivity values by adding Eqs. (4.12) and (4.13) into Eq. (4.9) and performing the second step fitting with the corresponding experimental data of the thin film/SiO_2 structures. The effect of introducing MoS_2 with increasing

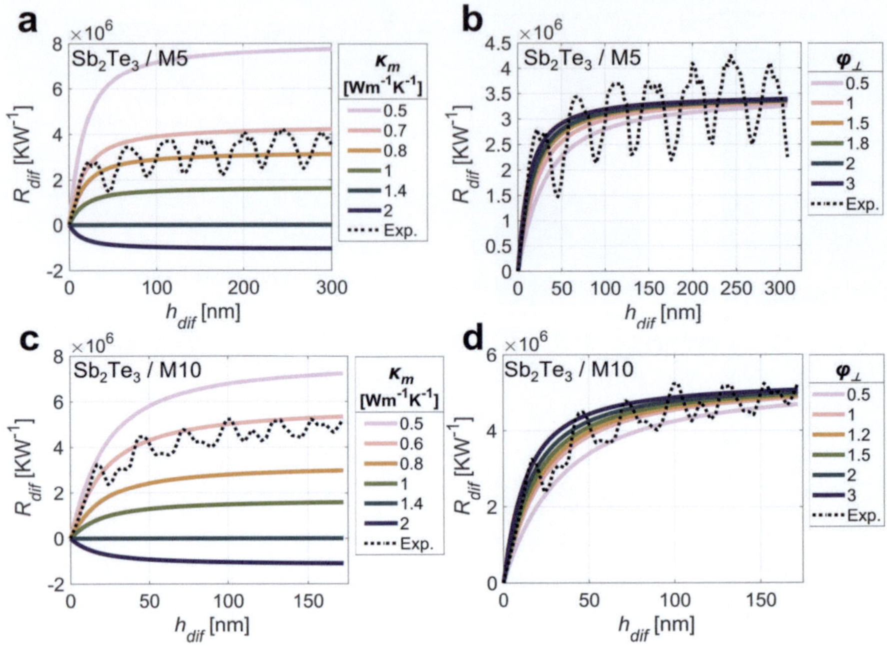

Fig. 5.4 Comparison of simulated $R_{dif}(h_{dif})$ dependences according to the analytical model (coloured lines) versus the experimental data (black dotted lines) for the anisotropic parameters of the multilayer thin films. **a**, **c** Varying κ_m and **b**, **d** varying $\varphi_\perp$

Table 5.2 Heat transport and TE parameters of the analysed samples at 316 K

	Sb_2Te_3	Sb_2Te_3/M5	Sb_2Te_3/M10
a_c (nm)	38.57 ± 1.35	38.17 ± 1.25	39.99 ± 2.77
α_c	8.19 ± 0.34	7.48 ± 0.28	7.65 ± 0.64
r_{int} (km^2 W^{-1})	–	2.19 × 10^{-9}	9.65 × 10^{-10}
κ_{xy} (W m^{-1} K^{-1})	4.50 ± 0.62 (3ω-method)	1.05 ± 0.43	0.69 ± 0.13
κ_z (W m^{-1} K^{-1})	1.52 ± 0.15 (3ω-method)	0.57 ± 0.22	0.58 ± 0.09
κ_e (W m^{-1} K^{-1})	0.46 ± 0.02	0.27 ± 0.01	0.14 ± 0.01
κ_L (W m^{-1} K^{-1})	4.06 ± 0.64	0.78 ± 0.44	0.55 ± 0.14
ZT	0.15 ± 0.05	0.88 ± 0.34	2.26 ± 0.62

thickness in the structure is related to a decrease of the anisotropic ratio ($\varphi_\perp$), also compared to the pristine sample 3ω-method values.

The oscillations observed in the experimental data in Fig. 5.4 are attributed to the different thermal conductivities of the individual Sb_2Te_3 and MoS_2 layers and their varied interfacial thermal insulances. However, as the target results are the thermal conductivities of the entire multilayer films, the average of such oscillations can be fitted with the “Muzychka-Spièce” model [14].

Their in-plane electronic and phononic thermal conductivities can also be obtained by employing Eq. (4.14) and $\kappa_L = \kappa_{xy} - \kappa_e$, respectively. The pristine sample has a higher electronic heat conductivity due to its enhanced electrical conductivity compared to the multilayer samples. Also, the lattice thermal conductivity is observed to decrease with the introduction of MoS_2 in the systems since the phonon scattering increases at the interfaces. Such scattering mechanisms could be of different nature, i.e., Umklapp, phonon/grain boundary, phonon/interface, point defect and phonon/ electron scattering, ultimately reducing the thermal conductivity values [2, 17, 18]. Other factors such as phase transitions, doping, alloying or the reduced thermal conductivity of few-layer MoS_2 can diminish the overall thermal conductivity (also studied in Sect. 5.3.2) [19–22].

Lastly, it is possible to calculate the final in-plane *ZT* of each sample (see Table 5.2). A record-high value of 2.26 ± 0.62 at 316 K is achieved in the Sb_2Te_3/M10 structure, increasing the efficiency by ~ 15 times compared to the pristine sample. This higher TE efficiency originates from the role of the 2D–3D interfaces, filtering the low energy charge carriers and promoting a transport channel through the thin MoS_2 and the topologically insulator Sb_2Te_3 layers as a quantum confinement effect, enhancing the power factor [8]. Simultaneously, the thermal conductivity values are reduced by the scattering effects at the interfaces. Overall, it is clear that controlling the thickness of stacked layers as well as their composition influences the transport through the created interfaces with the possibility of greatly enhancing the TE "Figure of Merit", providing a new direction for fine-tuning the electron and phonon transport of multilayer samples.

5.3 Silver/Antimony Telluride Nanocomposite

5.3.1 Electronic Transport Characteristics

To fully assess the influence of silver doping in the TE properties of Sb_2Te_3-based structures, five samples with different silver concentrations are fabricated. The silver atomic percentage is varied from 0% in the pristine Sb_2Te_3 sample (ST) to 0.90, 1.70, 5.90 and 7% in the AST1, AST2, AST3 and AST4 samples, respectively. Synthesis of such nanocomposite thin films over standard SiO_2/silicon substrates is conducted at 250 °C via the same radiofrequency sputtering technique described in the previous section, with an additional 30 min annealing at the same temperature in an inert atmosphere. In this case, the P_{RF} for the Sb_2Te_3 deposition is 40 W, while the silver power varies from 10 to 20 W, depending on the desired concentration, and $p_{chamber}$ is 2×10^{-3} mbar.

The composition characterisation of the samples (Sect. A.2.5) demonstrates their ternary nanocomposite structure, where distinct nanoscale phases could be observed in the ternary stoichiometry, involving three elements. Afterwards, Hall and experimental TE measurements are conducted in such five Sb_2Te_3-based structures

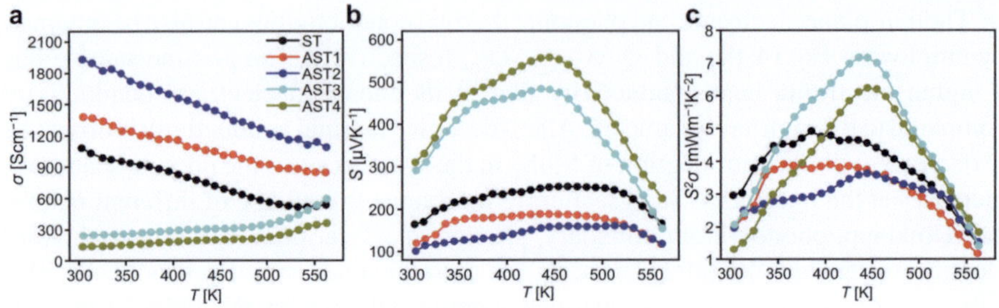

Fig. 5.5 In-plane electronic parameters of the nanocomposite samples. **a** Electrical conductivities, **b** seebeck coefficients and **c** power factors against temperature

employing the same equipment described in Sect. 5.2.2. Figure 5.5a shows an increase in the electrical conductivity for the two initial doping concentrations compared to the pristine sample, however, at higher doping, the ST sample has higher conductivity than both the AST3 and AST4 samples. Interestingly, the relation of the conductivity with temperature changes from metallic (decreasing slope at ST, AST1 and AST2) to semiconductor behaviour (increasing slope at AST3 and AST4) as the concentration surpasses 1.70%. Such phenomenon is attributed to a hopping-dominated conduction mechanism in the highly doped nanocomposites [23]. The enhancement of the Seebeck coefficient (Fig. 5.5b) compensates for the overall decreased electrical conductivities of AST3 and AST4, resulting in high power factors at 435 K (7.17 ± 0.70 m W m^{-1} K^{-2}), as shown in Fig. 5.5c. Similarly to the superlattices, S tends to decrease at higher temperatures due to the combined charge transport of holes and electrons, also occurring in $AgSbTe_2$ [24]. However, the majority of the charge transport is still dominated by holes, as the Seebeck coefficients remain positive.

Table 5.3 shows the quantitative results of the Hall and thermoelectric measurements at 303 K, with an increase of the carrier density of up to 4.47×10^{19} cm^{-3} proportional to the doping and a following decrease in the AST3 and AST4 samples. The initial increase in carrier density is ascribed to the formation of silver substitutional defects during the growth process, which act as acceptor dopants in low silver concentrations, hence, increasing the charge carrier conductivity [25].

Moreover, the typical TE inverse relation of carrier density versus mobility is lost at the AST3 and AST4 samples, with a proportional decrease of the mobility when the concentration reduces, causing an anomalous Seebeck coefficient. This reduced mobility trend cannot be explained only by carrier filtering, quantum confinement or ionised scattering since, in such cases, only a slight negative impact is expected on the mobility [26]. Thus, to understand the decreased-mobility mechanisms, the values of the DOS effective mass ($m_{DOS,eff}$) are calculated and compared for the different silver concentrations following Eq. (2.18).

The obtained effective masses are also shown in Table 5.3. The larger DOS masses in the AST3 and AST4 samples indicate the non-trivial role that such parameter plays in the S enhancement [27]. Additionally, the TE properties are improved from the formation of a potential barrier between the two phases in the ternary

Table 5.3 In-plane electronic transport parameters of the nanocomposites at 303 K

	ST	AST1	AST2	AST3	AST4
n × 10^{19} (cm^{-3})	1.12	2.08	4.47	0.89	0.72
μ (cm^2 V^{-1} s^{-1})	603.96	413.75	269.90	170.60	118.20
$m_{DOS,eff}$ (kg)	3.71×10^{-31}	4.09×10^{-31}	5.71×10^{-31}	5.62×10^{-31}	5.20×10^{-31}
σ (Scm^{-1})	1083.65	1378.69	1932.79	242.97	135.97
S (μVK^{-1})	165.33 ± 11	120.59 ± 8	101.02 ± 7	291.65 ± 20	310.80 ± 22
$S^2\sigma$ (m W m^{-1} K^{-2})	2.96 ± 0.23	2.01 ± 0.16	1.97 ± 0.15	2.07 ± 0.16	1.32 ± 0.10

nanocomposites (see Sect. A.2.5), equivalent to the superlattice carrier filtering mechanism. The number of interfaces between such two phases increases with the carrier concentration, resulting in intensified carrier filtering. However, this also diminishes the number of transported charge carriers, hence decreasing the detected carrier concentration and electrical conductivity.

5.3.2 *Heat Transport Measurements*

To correctly define *ZT*, the heat contribution to the TE properties has to be appropriately quantified. The xSThM methodology is also applied here for such purpose, hence the samples have to be post-treated with BEXP to reveal their inner surface and be able to scan it with HV SThM, resulting in topography and thermal signal data similar to the one shown in Sect. 5.2.3 (see Sect. A.3).

The results of the corresponding fittings are expressed in Table 5.4. The goodness of fit can be evaluated for all the anisotropic layers, having an individual adjusted R^2 value > 0.99 for all the samples, proving that the fitting results are accurate. Also, multiple measurements are performed at different points of the materials to eliminate the effect of local variation of defect density on the surfaces, all of them providing equivalent results. Low values for the overall thermal conductivities are obtained in all the nanocomposites, indicating a clear decrease of the heat transport when silver is incorporated in a pristine Sb_2Te_3 system due to combined phonon scattering in both the in-plane and cross-plane directions.

As explained in Sect. A.2.5, the crystallinity of the layers diminishes with increased silver concentration, also indicating a reduction of the crystal domains. In order to decrease the crystallinity, defects are introduced in the system to perturb the structure periodicity, acting as crystal nucleation sources due to high interfacial energy and weak bonding. Silver nucleation is likely to form given the greater electronegativity difference between silver and tellurium and silver and antimony compared to the difference between antimony and tellurium. Thus, one can say

Table 5.4 Heat transport and TE parameters of the nanocomposites at 303 K

	AST1	AST2	AST3	AST4
a_c (nm)	49.07 ± 1.02	52.55 ± 0.66	53.26 ± 0.87	52.82 ± 2.04
α_c	7.94 ± 0.15	1.88 ± 0.02	5.25 ± 0.08	6.96 ± 0.27
r_{int} (km^2 W^{-1})	1.12×10^{-9}	6.40×10^{-10}	3×10^{-8}	1.96×10^{-10}
κ_{xy} (W m^{-1} K^{-1})	0.92 ± 0.02	0.80 ± 0.03	1.03 ± 0.09	0.89 ± 0.02
κ_z (W m^{-1} K^{-1})	0.22 ± 0.01	0.23 ± 0.01	0.11 ± 0.01	0.23 ± 0.01
ZT	0.66 ± 0.09	0.75 ± 0.12	0.61 ± 0.12	0.45 ± 0.07

that a higher density of defects (silver substitutions) forms more nucleation centres, ultimately decreasing the grain size. As seen in the superlattice thermal transport, such type of interface promotes mid-to-long wavelengths phonon scattering mechanisms, inhibiting the phonon propagation and minimising the thermal conductivity. Additionally, the high concentration of silver domains is associated with structural distortions, such as lattice dislocations, with possible strain relaxation domains occurring during the growth process [28]. These atomic-scale distortions scatter short wavelengths phonons, ensuring the total spectrum reduction of the heat transport.

From the superlattice study, it is clear that the decrease in thermal conductivity corresponds to an enhanced "Figure of Merit". However, the combination of the low heat transport in the nanocomposites with the calculated power factors is not enough to achieve suitable *ZT* values at RT, as seen in Table 5.4. This situation can be overcome by assuming a constant thermal conductivity of such thin films as the temperature increases, allowing to use bigger power factor values still in the range of moderate temperature. This assumption is a typical trend for phonon-dominated thermal conductivity systems, such as the ones reported here, and has been widely employed elsewhere [28, 29], with the only exception being bipolar heat transport (not observed in our system until 380 K).

With that, one can achieve higher TE efficiency values of 2.03 and 1.64 at 375 K on the AST3 and AST4 samples, which, to the best of the author's knowledge, are the highest reported values in silver-based thermoelectric nanocomposites for this temperature range.

5.4 Summary

Two different methodologies have been proposed and studied in this chapter with the goal of achieving elevated thermoelectric efficiencies in layered structures. A detailed examination of the structural, electronic and thermal characteristics of multilayer and nanocomposite Sb_2Te_3-based materials has been conducted for the complete assessment of the thermoelectric feasibility of such structures.

Enhanced power factors have been obtained due to the selective screening of charge carriers through the interfaces. These interfaces also promote the reduction of the thermal transport in the anisotropic layers following its phonon scattering mechanisms, a beneficial tool for enhanced thermoelectric energy conversion. The direction-dependent thermal transport quantification is easily and accurately accessible by employing the xSThM analytical method, capable of measuring both in-plane and cross-plane thermal conductivities. Moreover, both the electronic and phononic transport are engineered by the modification of the crystalline orientation and grain size as well as the incorporation of low thermal conductivity MoS_2 layers in the case of the superlattice structures. Specifically, the introduction of thin film MoS_2 has been found to greatly enhance the Seebeck coefficient and simultaneously decrease the thermal conductivity, achieving a value of $ZT = 2.26 \pm 0.62$ at 316 K for a Sb_2Te_3/M10 multilayer sample. Also, doping with silver has been found beneficial to increase the ZT values at higher temperatures. This study highlights the direct influence and importance of the interfaces between 2D–3D materials as well as the precise selection of their composition as tunable building blocks capable of independent modulation of the electronic and phononic transport, leading to feasible platforms for enhanced thermoelectric energy management.

References

1. Snyder, G. J., & Toberer, E. S. (2008). Complex thermoelectric materials. *Nature Materials, 7*, 105–114.
2. Caylor, J. C., Coonley, K., Stuart, J., Colpitts, T., & Venkatasubramanian, R. (2005). Enhanced thermoelectric performance in PbTe-based superlattice structures from reduction of lattice thermal conductivity. *Applied Physics Letters, 87*, 023105.
3. Xie, W., Tang, X., Yan, Y., Zhang, Q., & Tritt, T. M. (2009). Unique nanostructures and enhanced thermoelectric performance of melt-spun BiSbTe alloys. *Applied Physics Letters, 94*, 102111.
4. Venkatasubramanian, R., Siivola, E., Colpitts, T., & O'Quinn, B. (2001). Thin-film thermoelectric devices with high room-temperature figures of merit. *Nature, 413*, 597–602.
5. Ahmad, M., Agarwal, K., Kumari, N., & Mehta, B. R. (2017). KPFM based investigation on the nature of Sb_2Te_3:MoS_2 and Bi_2Te_3:MoS_2 2D interfaces and its effect on the electrical and thermoelectric properties. *Applied Physics Letters, 111*, 023904.
6. Ye, L.-H., et al. (2008). First-principles study of the electronic, optical, and lattice vibrational properties of $AgSbTe_2$. *Physical Review B, 77*, 245203.
7. Gao, Y., He, Y., & Zhu, L. (2010). Impact of grain size on the Seebeck coefficient of bulk polycrystalline thermoelectric materials. *Chinese Science Bulletin, 55*, 16–21.
8. Jiang, Y., et al. (2012). Landau quantization and the thickness limit of topological insulator thin films of Sb_2Te_3. *Physical Review Letters, 108*, 016401.
9. Shi, X., et al. (2017). An analytical Seebeck coefficient model for disordered organic semiconductors. *Physics Letters A, 381*, 3441–3444.
10. Fang, B., Zeng, Z., Yan, X., & Hu, Z. (2013). Effects of annealing on thermoelectric properties of Sb_2Te_3 thin films prepared by radio frequency magnetron sputtering. *Journal of Materials Science: Materials in Electronics, 24*, 1105–1111.
11. Lee, K. H., et al. (2020). Nanoparticles in $Bi_{0.5}Sb_{1.5}Te_3$: A prerequisite defect structure to scatter the mid-wavelength phonons between Rayleigh and geometry scatterings. *Acta Materialia, 185*, 271–278.
12. Ahmad, M., Varandani, D., & Mehta, B. R. (2018). Large surface charge accumulation in 2D MoS_2/Sb_2Te_3 junction and its effect on junction properties: KPFM based study. *Applied Physics Letters, 113*, 141603.

13. Rieger, F., et al. (2018). Low intrinsic c-axis thermal conductivity in PVD grown epitaxial Sb_2Te_3 films. *Journal of Applied Physics, 123*, 175108.
14. Spièce, J., et al. (2021). Quantifying thermal transport in buried semiconductor nanostructures via cross-sectional scanning thermal microscopy. *Nanoscale, 13*, 10829–10836.
15. Zhu, W., Zheng, G., Cao, S., & He, H. (2018). Thermal conductivity of amorphous SiO_2 thin film: A molecular dynamics study. *Science and Reports, 8*, 10537.
16. Shanks, H. R., Maycock, P. D., Sidles, P. H., & Danielson, G. C. (1963). Thermal conductivity of silicon from 300 to 1400°K. *Physical Review, 130*, 1743–1748.
17. Liu, Z., et al. (2020). High power factor and enhanced thermoelectric performance in Sc and Bi codoped GeTe: Insights into the hidden role of rhombohedral distortion degree. *Advanced Energy Materials, 10*, 2002588.
18. Zhang, Q., et al. (2015). Enhancement of thermoelectric performance of n-type PbSe by Cr doping with optimized carrier concentration. *Advanced Energy Materials, 5*, 1401977.
19. Zhou, W.-X., et al. (2020). Thermal conductivity of amorphous materials. *Advanced Functional Materials, 30*, 1903829.
20. Suwardi, A., et al. (2020). Tailoring the phase transition temperature to achieve high-performance cubic GeTe-based thermoelectrics. *Journal of Materials Chemistry A, 8*, 18880–18890.
21. Suwardi, A., et al. (2020). Achieving high thermoelectric quality factor toward high figure of merit in GeTe. *Materials Today Physics, 14*, 100239.
22. Gu, X., Li, B., & Yang, R. (2016). Layer thickness-dependent phonon properties and thermal conductivity of MoS_2. *Journal of Applied Physics, 119*, 085106.
23. Agarwal, K., & Mehta, B. R. (2014). Structural, electrical, and thermoelectric properties of bismuth telluride: Silicon/carbon nanocomposites thin films. *Journal of Applied Physics, 116*, 083518.
24. Wolfe, R., Wernick, J. H., & Haszko, S. E. (2004). Anomalous hall effect in $AgSbTe_2$. *Journal of Applied Physics, 31*, 1959–1964.
25. Navrátil, J., Klichová, I., Karamazov, S., Šrámková, J., & Horák, J. (1998). Behavior of Ag admixtures in Sb_2Te_3 and Bi_2Te_3 single crystals. *Journal of Solid State Chemistry, 140*, 29–37.
26. Wiendlocha, B., Misra, S., Dauscher, A., Lenoir, B., & Candolfi, C. (2021). Residual resistivity as an independent indicator of resonant levels in semiconductors. *Materials Horizons, 8*, 1735–1743.
27. Namiki, H., et al. (2022). Relationship between the density of states effective mass and carrier concentration of thermoelectric phosphide $Ag_6Ge_{10}P_{12}$ with strong mechanical robustness. *Materials Today Sustainability, 18*, 100116.
28. Ferrer-Argemi, L., et al. (2019). Silver content dependent thermal conductivity and thermoelectric properties of electrodeposited antimony telluride thin films. *Science and Reports, 9*, 9242.
29. Yu, Z., et al. (2018). Phase-dependent thermal conductivity of electrodeposited antimony telluride films. *Journal of Materials Chemistry C, 6*, 3410–3416.

Chapter 6
Local Thermoelectricity in Layered Nanostructures

Abstract This chapter explores the local control of thermoelectric properties in layered nanostructures, with a particular emphasis on the Seebeck coefficient. The goal of this chapter is to demonstrate how nanoscale modifications, specifically nanoconstrictions, can be used to spatially tune thermovoltage generation and enable in situ characterisation of thermoelectric (TE) behaviour. The chapter introduces scanning thermal gate microscopy (STGM) as a central technique, allowing precise probing of local thermovoltage responses. After describing the experimental setup and the underlying conversion principles, the focus shifts to two types of patterned nanoconstrictions: bow tie geometries in encapsulated graphene and hole arrays in tin diselenide. These engineered structures enable the manipulation of local electronic and thermal transport, directly impacting the Seebeck coefficient and the resulting thermoelectric response. The final section addresses thickness-induced thermoelectric boundaries in graphite, further illustrating how structural parameters can be harnessed to control TE behaviour. Together, these studies highlight the potential of 2D materials for nanoscale thermoelectric applications and establish a methodology for achieving spatial control over energy conversion at the device level. The techniques and results presented here lay the foundation for future strategies in designing thermoelectric platforms with locally optimised performance.

6.1 Introduction

Having understood the role of thermal transport and its anisotropy in energy conversion applications and the nature of the overall TE performance enhancement in 2D-based systems, the contribution of the power factor to the "Figure of Merit" formula is the only remaining unknown. Specifically, the Seebeck coefficient, also known as thermopower, holds an essential influence on such efficiency calculation, given that it is the parameter that quantifies the relation between the induced TE voltage (thermovoltage) and the applied temperature difference within the material, according to the Seebeck effect (see Sect. 2.5.2) [1]. For this reason, it is crucial to not only understand the behaviour of the Seebeck coefficient in 2D structures but also to engineer

S. Gonzalez-Munoz, *Interplay of Nanoscale Physical Properties in 2D Materials and Heterostructures*, Springer Theses, https://doi.org/10.1007/978-3-032-14271-9_6

suitable approaches to locally modulate and control it. In the previous chapter, the two methodologies discussed were capable of achieving a notorious enhancement of the power factor based on the Seebeck modulation along the entire two-dimensional structures by the effect of the interfaces in the electronic transport, however, the question of direct local Seebeck coefficient modification at a nm scale in the in-plane direction remains unsolved. Such spatially modulated energy management could be beneficial for the implementation of single vdW materials as thermocouples or accessible heat dissipation systems.

It is clear that the reduction in the dimensions of TE systems leads to an engineered Seebeck coefficient and a potentially increased efficiency. Moreover, localised spatial phenomena such as defects or interfaces with metallic electrodes drastically affect the TE performance of the device [2, 3]. Following these reasonings, Harzheim et al. recently developed an approach to monitor and manipulate the spatial dependence of the local Seebeck coefficient by mapping nanoconstrictions in two-dimensional structures with an STGM setup [1, 4]. They attributed the variation of the Seebeck signal to the restricted effective electron mean free path due to edge scattering at the nanoconstriction area.

Here, an improved STGM system is employed to analyse three types of nanoconstrictions in an h-BN/graphene/h-BN encapsulated heterostructure and a tin diselenide ($SnSe_2$) flake for the first time. The variation in the geometries of the constrictions and the materials' composition provides an expanded insight into how these two factors affect the local TE behaviour. The STGM technique (see Sect. 2.4.4) probes the local variations of the thermovoltage (hence, the Seebeck coefficient) non-destructively in vdW structures with nanoscale resolution. This opens a possibility for the precise deconvolution of the layer's thermoelectric signal from other structural effects, e.g., the electrode contribution, the scattering impact or the thickness variation, which is also investigated at the end of the chapter on a thin graphite flake. Also, the analysis of 2D structures allows the use of a backgate to tune their electronic characteristics. The effect of this gating phenomenon is studied in the encapsulated structure, where different values of doping lead to modulated thermovoltage.

The contributions of Dr. Alexandros El Sachat, Dr. Eli G. Castanon, Dr. Charalambos Evangeli and Dr. Peng Xiao must be acknowledged for this chapter. The bow tie heterostructure was fabricated and patterned by Dr. Eli G. Castanon at the National Graphene Institute and the hole arrays post-TMD sample was fabricated and patterned by Dr. Alexandros El Sachat and Dr. Peng Xiao at the Catalan Institute of Nanoscience and Nanotechnology. The characterisation and results interpretation of both patterned samples and the graphite device were performed by myself in collaboration with Dr. Charalambos Evangeli, Dr. Alexandros El Sachat and Dr. Eli G. Castanon in the Lancaster University Physics Department.

6.2 Scanning Thermal Gate Microscopy for Thermovoltage Studies

6.2.1 Thermoelectric Setup

In order to measure the thermoelectric response of the layered materials, the HV SPM setup employed in Chap. 4 to map the SThM thermal signal is accordingly modified to provide a bias capable of heating the probe and reading the resulting thermovoltage difference from the electrodes of the device, representing a novel configuration known as scanning thermal gate microscopy (STGM) (Fig. 6.1).

Specifically, a Bruker Anasys Instruments Therma-Lever™ AN2-200 two-legged doped silicon probe is powered by a GEN (Keithley 3390), providing a high AC bias of 18 V at a frequency of 61 Hz, controlled through the BRG (corresponding to 3.60 V at the probe). Such driving excitation increases the temperature of the probe by tens of K above the reference system temperature, causing a thermal excess concentration at the tip's apex, hence, acting as a local heater while being in contact with the device. This local temperature rise in the heating spot generates a voltage difference in the measured material arising from the Seebeck principle. As mentioned before, the resulting potential is known as thermovoltage (V_{TE}), which is directly related to the spatial distribution of the Seebeck coefficient of the layered structure and, therefore, its TE properties. The drain and source electrodes track such thermovoltage along all the mapped 2D space while the heating source changes its position. Both electrodes are connected to a breakout box (BKB), acquiring the bias drop and introducing it into the FLT (Stanford Research Systems SR650) to condition the TE signal. Afterwards,

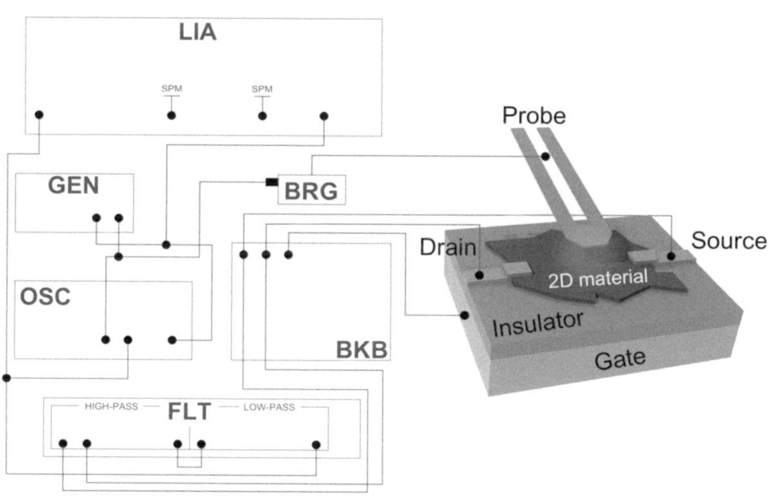

Fig. 6.1 STGM custom-built setup and analysed device structure (GEN, BRG, OSC, LIA, BKB and FLT are the generator, bridge, oscilloscope, lock-in amplifier, breakout box and filter, respectively)

a LIA (Stanford Research Systems SR830 DSP) is used to amplify and monitor the in-phase and out-of-phase response sent to the SPM controller and shown as thermovoltage maps with nanoscale spatial resolution.

The system is based on the regular AFM configuration with the laser tracking of the cantilever's deflection, allowing to acquire topographical maps simultaneously with the thermovoltage scans. Moreover, STGM does not require an electrical contact between the probe and the sample, allowing to consider the tip as a thermal gate which does not introduce any external current into the device in equilibrium. The samples analysed in this chapter are layered materials deposited on standard SiO_2 films grown on crystalline conductive silicon substrates. Hence, by adding an electrical connection to the silicon layer, the device can function as a field-effect transistor capable of precise modulation of the doping levels in the studied vdW materials in non-equilibrium conditions. One of the electrodes functions as an electric source, generating a current flow through the 2D material (channel) that reaches the other electrode (drain). This electronic signal can be modulated by the electric field generated via the potential difference between the backgate and the drain, with zero gate–drain current due to the separating insulator.

Similarly to the setup in Chap. 4, it is crucial to ensure a singular conductive heat transfer pathway through the apex of the tip to the vdW material to maintain a nanoscale resolution of the TE properties. Thus, the measurements are conducted under HV environment ($< 2 \times 10^{-6}$ mbar) to prevent unwanted thermal losses and surface contamination. Additionally, the sample is thermally bonded on the copper stage, where the reference temperature is monitored. This stage is thermally isolated from the rest of the microscope frame via low heat conduction glass spacers and incorporates a copper braid connected to a liquid nitrogen reservoir, allowing the cooling of the sample to low temperatures (down to 120 K).

6.2.2 Thermoelectric Conversion Principle

The explained local probe microscopy configuration registers the TE voltage generated in the in-plane direction of the studied structure. However, to transform said thermovoltage and directly quantify the Seebeck coefficient ($S(\xi)$) of the local area in the material, the following equation can be used [4]:

$$V_{TE}(x) = -\int_{x_d}^{x_s} S(\xi)\frac{\partial T(\xi - x)}{\partial \xi}d\xi \tag{6.1}$$

where ∂T is the tip/electrodes temperature excess, ξ is the x-direction integration variable and x, x_s and x_d are the tip's, source's and drain's positions, respectively. This expression illustrates how the change of the Seebeck coefficient at different TE behaviour areas influences the respective thermovoltages after a position-dependent temperature gradient has been created ($\partial T \partial \xi^{-1}$). If the excess temperature function is assumed to be spatially symmetric, the asymmetric quantification of V_{TE} and S around the measurement position can be achieved, accounting for their local modulation at length scales determined by the heat resistances of the sample. The local asymmetry of the Seebeck, reflected with a spatial gradient, can therefore be translated in a thermovoltage different than zero, e.g., at the sides of a junction of two materials with different Seebeck behaviour. However, if the S is kept constant, no spatial gradient is created, and the thermovoltage value is nullified.

6.3 Patterned Nanoconstrictions

6.3.1 Geometrical Bow Tie Graphene

Device Fabrication and Preliminary Characterisation

In the previous STGM studies performed on graphene nanoconstrictions [1, 4], the layered material was directly deposited on top of a SiO_2 layer, leading to the positive doping of graphene via large potential fluctuations originating on the insulator's layer impurities and charged surface states [5]. Furthermore, SiO_2 can induce electron/hole puddles in the vicinity of the Dirac point [6]. The overall SiO_2 parasitic modulation of the electronic properties of graphene can cause a diminished Seebeck coefficient [7]. Moreover, if the employed tip is electrically conductive, a charge transfer could arise from the probe to the graphene layer, varying the doping state of the vdW material. Thus, an encapsulation of the graphene layer with insulating h-BN is required to provide a flat, matching and clean interface, reducing surface roughness, contaminants and charge traps and mitigate the electric interaction with the tip, ultimately preserving graphene's intrinsic properties. Lastly, asymmetric patterned constrictions with multiple widths must be compared to their studied symmetric counterparts to account for the geometrical influence in the TE properties of the heterostructure.

To encapsulate the graphene between the h-BN layers, the heterostructure fabrication methodology explained in Sect. 3.3 is employed, changing the carrier material from PDMS to PMMA to visually distinguish the h-BN layers. First, a large-sized h-BN flake is deposited on a pretreated 90 nm thick SiO_2/silicon substrate avoiding any macroscopic defects. Afterwards, the ~ 3 nm thin top h-BN layer is exfoliated and used to pick-up the 1LG, covering it entirely. The encapsulation is completed when the 1LG and top h-BN are deposited on the bottom h-BN flake, completely isolating the graphene (Fig. 6.2a and b).

To pattern the heterostructure, EBL is employed (see Sect. 2.3.4). First, the shape design is conducted with LayoutEditor so that the final structure contains two external half bow tie/rectangular junctions and a centre narrow strip transition (Fig. 6.2d) and fitted over the graphene-delimited area with the aid of alignment marks (Fig. 6.2c). Then, the PMMA positive resist is deposited ensuring the covering of the entire heterostructure and the EBL exposure is performed outside of the design stipulated

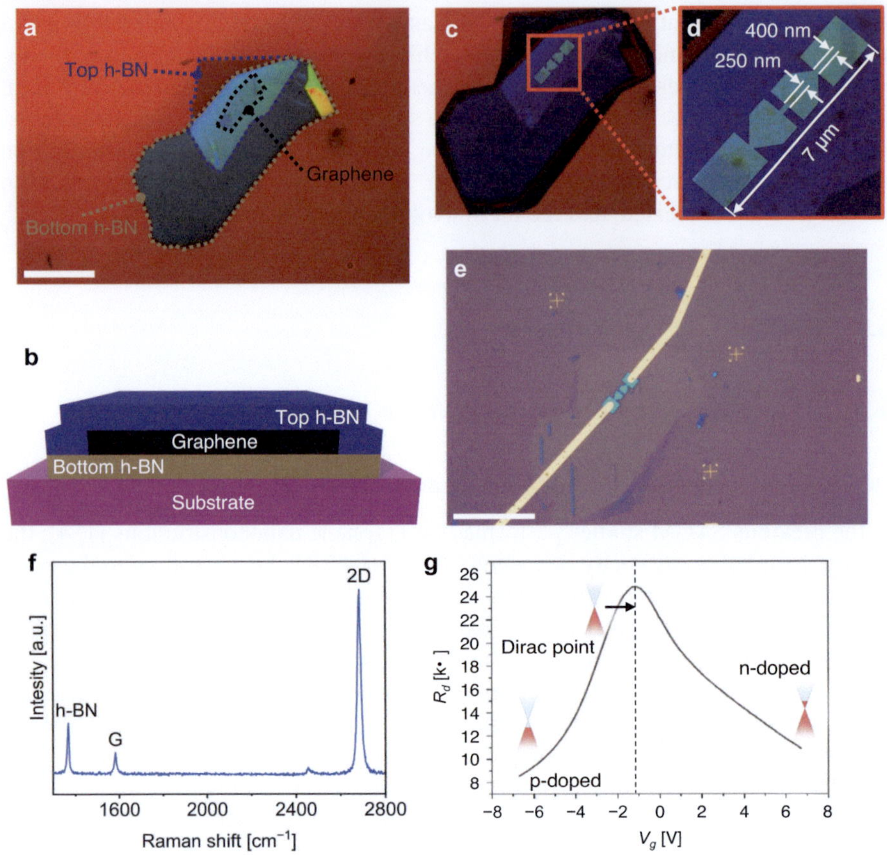

Fig. 6.2 Preliminary characterisation of the heterostructure. **a** Contrast-enhanced optical image of the h-BN/graphene/h-BN heterostructure on the SiO_2/silicon substrate (scale bar is 20 μm), **b** scheme of the encapsulated heterostructure, **c** positioning and design of the patterned structure over the layers before the lithography, **d** magnified view of the pattern with the total length and constrictions' widths dimensions, **e** optical image of the fabricated pattern with the deposited metallic electrodes (scale bar is 30 μm), **f** Raman spectrum of the patterned heterostructure and **g** device resistance (R_d) versus gate voltage (V_g) curve showing the different doping regions

area. The subsequent development of the resist and RIE of the exposed layers result in the light blue contrast patterned heterostructure seen in Fig. 6.2e. This figure also shows the deposited gold (~ 50 nm)/chromium (~ 1–5 nm) contacts working as electrodes. These are fabricated following the process depicted in Sect. 3.4.2, contacting the side of the graphene layer and, hence, forming an ohmic edge contact configuration in a single dimension. The previous STGM study has also shown that a top contact configuration could lead to a backgate shielding of the graphene region underneath the gold electrode, meaning that external charge doping from the metal would be transferred to the vdW material, affecting the thermovoltage values [4]. This parasitic phenomenon is avoided here by protecting the graphene with h-BN encapsulation from a surface-type contact, ensuring a 1D electrical junction without any other interaction between the layered structure and the electrodes. The pads of such electrodes are finally wire-bonded to external connections to finish the fabrication.

The composition of the heterostructure is characterised by Raman spectroscopy before the patterning (Fig. 6.2f) employing a Renishaw® inVia™ confocal Raman microscope with an excitation source of 532 nm of wavelength. The resulting spectrum shows the characteristic graphene G and 2D peaks, with their ratio of intensities (> 2) and Raman shift positions (1582 and 2683 cm^{-1}, respectively) denoting a monolayer thickness. Also, the h-BN layers are represented by a symmetric peak at 1366 cm^{-1} [8, 9]. The finished device's electrical response is also measured with the electrodes as a two-probe system. In this case, a first LIA provides the driving current at an output voltage of 0.10 V and 2.70 kHz through two resistances connected in series with the device. The first of these resistances (1 MΩ) restricts the current flowing through the device to a constant value of 1 μA and the second one (10 kΩ) monitors such current via a second LIA. Such configuration allows to acquire a total resistance value for the heterostructure, the electrodes and their contact interface of ~ 22 kΩ. Backgate DC voltage is provided from a Keithley Sub-Femtoamp Remote SourceMeter® SMU Instrument to quantify the gating influence on the device's resistance (R_d) as shown in Fig. 6.2g. This measurement is conducted under HV at RT, finding the Dirac point at ≈ -1.16 V (representing a slight n-type doping in the graphene layer).

Due to the large voltage supplied to the probe during the STGM measurements, vibrations are created along the length of the cantilever, hence, the topographical maps would lack a suitable quality and resolution, requiring further image postprocessing. For this reason, prior HV AFM and SThM scans are conducted over the area of interest, extracting their profiles to quantify the thickness of the elements and their thermal properties (Fig. 6.3). The thickness of the electrodes and heterostructure is revealed as 45 and 25 nm, respectively. Also from the AFM map, the widths of the half bow tie and centre constrictions are measured to be ~ 370 and 240 nm, respectively, in agreement with the patterning design. The thermal signal SThM map shows a darker contrast difference between the heterostructure compared to the surrounding SiO_2 and gold surfaces. As discussed in Chap. 4, this is indicative of a higher thermal conductivity in the encapsulated structure, which is expected due to the large in-plane heat transport values of both graphene and h-BN. The combined thermal conductivity value of the entire encapsulated specimen has also been measured by Karak et al. [10], specifying a value of 850 W m^{-1} K^{-1}, which is larger than both the SiO_2 and gold heat conductivities (1.40 and 314 W m^{-1} K^{-1}, respectively) [11, 12]. Thermal scattering can also be observed at the boundaries of the heterostructure and the limits of the peeled regions of the top h-BN layer as a higher thermal signal arising from diminished heat conductivity, also identified from the extracted profiles. The probe geometry can also have an effect in such higher thermal signal voltages, requiring further modelling to deconvolute the tip's contribution from the edge scattering of the device out of the scope of the TE measurements described in this chapter.

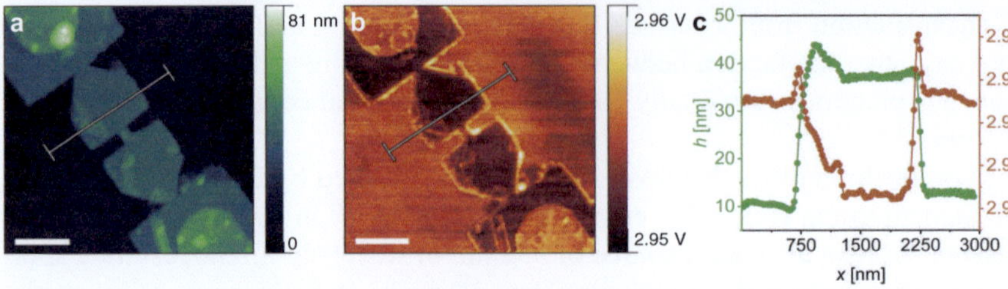

Fig. 6.3 HV SThM analysis of the device. **a** Topography and **b** thermal signal maps (scale bars are 1 μm) and **c** corresponding extracted profiles

Thermoelectric Response of the Encapsulated Heterostructure

Harzheim et al. proved that the modification of the Seebeck coefficient is directly related to the limitation of the electron mean free path through the constriction width [1]. This phenomenon arises from the scattering mechanisms that charge carriers face in layered structures, such as lattice defects and edge, grain boundary or surface scattering, however, the mean free path can also change with the carrier density or the temperature [13]. In pristine h-BN encapsulated graphene, the electronic mean free path is in the hundreds of nm to tens of μm range [14], however, the scattering caused at the close edges of a nm wide constriction like the ones described here drastically reduces such values. Therefore, as the edge scattering becomes more dominant, the mean free path follows a position dependence $\lambda(x)$ as [1]:

$$\lambda(x) = \lambda_0 \left[1 + \alpha_t \left(\frac{\lambda_0}{w(x)} \right)^{\alpha_s} \right]^{-1} \tag{6.2}$$

where λ_0 is the bulk mean free path of the structure, $w(x)$ is its position-dependent width and α_t and α_s are the numerical coefficients accounting for the transport mode and the scattering influence on the mean free path. Once the mean free path and constriction width are known, these two parameters can be related to the Seebeck coefficient and the thermovoltage employing Eq. (6.1) and the Mott's formula for metals [15]. As the graphene is shielded from parasitic S contributions by its encapsulation, the measured TE signal should be enhanced, nevertheless, the constrictions in the studied devices are expected to show a similar TE gradient as in the previous studies.

HV STGM scans are performed on the heterostructure at RT (~ 297 K) and with a probe excess temperature of approximately 50 K. Such heat excess can be monitored with a prior power calibration explained in Sect. 4.2, knowing the probe's temperature generated by a known supplied voltage. The backgate allows to control the doping level of the encapsulated graphene, obtaining the thermovoltage maps for both p-doped ($V_g = -6.80$ V) and n-doped ($V_g = 5.70$ V) configurations (Fig. 6.4a and b, respectively).

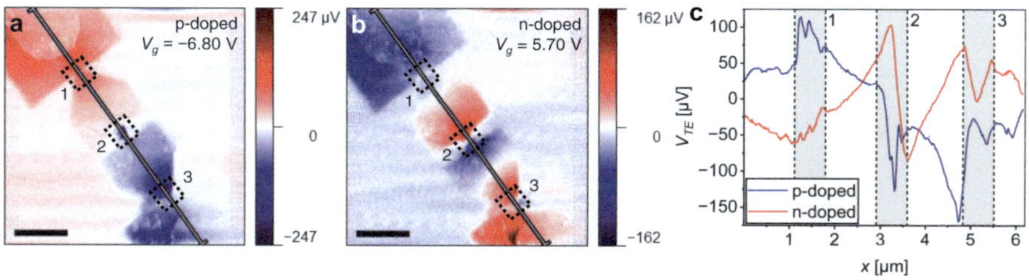

Fig. 6.4 RT HV STGM analysis of the device. Thermovoltage maps of the sample with induced **a** p-doped and **b** n-doped regimes (scale bars are 1 μm) and **c** corresponding extracted profiles with the constriction areas highlighted

By looking at the central constriction (area 2), one can observe a polarity inversion from positive to negative thermovoltage that occurs at both doping configurations. This phenomenon also happens at the asymmetric constrictions 1 and 3, but, this time, only at the n-doped setup. P-doped constrictions 1 and 3 show constant thermovoltage polarity given the higher absolute doping voltage ($|V_{g,p\text{-}doped}| - |V_{g,n\text{-}doped}| = 1.10$ V), which increases the positive charges in graphene (with respect to the negative charges at the n-doped configuration) shifting the thermovoltage away from zero at the asymmetric constrictions. Such reasoning could be verified by inverting the thermovoltage response of the p-doped structure and comparing it to the original n-doped counterpart. In this case, the thermovoltage at the electrodes would be the same (overlapping) for both doping scenarios since they are not affected by the gate voltage, however, the p-doped encapsulated graphene would show a shifted signal with the same slope trends compared to the n-doped counterpart. This shift would cause a modified zero-thermovoltage transition compared to the n-doped sample.

Profiles are also extracted from both STGM maps and represented in Fig. 6.4c. High relative thermovoltage shifts occur at both the centre and bottom constriction regions (areas 2 and 3, respectively), but the polarity is changed at each doping regime. Despite the two half bow tie constrictions theoretically having a similar geometric structure, the edges and defects (identifiable with a closer inspection of such areas at the AFM and SThM maps shown in Fig. 6.3) at area number 3 function as disorder scattering sources increasing the thermovoltage response compared to a more geometrically pristine area 1. As mentioned before, a higher degree of disorder corresponds to a lower electronic mean free path and a modified Seebeck coefficient, which, in turn, affects the thermovoltage shift. Such structural defects are caused by the tip's peeling of the top h-BN layers at regions 2 and 3 during previous scans, resulting in folds, ridges or cracks. Lastly, to account for the capacitive gating influence of the tip on the graphene layer (capable of causing unwanted charge accumulation on the vdW material), the same scans are repeated at the identical p- and n-doped regimes varying the supplied DC voltage to the tip from 3 to − 3 V. When the profiles are extracted, a minimal difference is observed, with the average deviation being a statistically acceptable 3%.

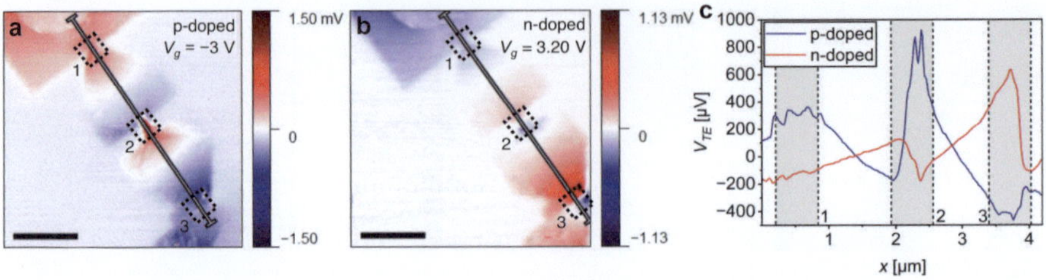

Fig. 6.5 LT HV STGM analysis of the device. Thermovoltage maps of the sample with induced **a** p-doped and **b** n-doped regimes (scale bars are 1 μm) and **c** corresponding extracted profiles with constriction areas highlighted

LT (~ 170 K) HV STGM measurements are conducted on the same device area to deconvolute the TE signal and achieve higher thermovoltage values. If the same amount of power is supplied to the probe, keeping a heat source similar to the RT measurements, the thermal excess between the tip and the nitrogen-cooled sample is transformed to ~ 177 K. The LT thermovoltage maps and corresponding profiles for each doping configuration are depicted in Fig. 6.5. This time, the gating potential's magnitudes are kept similar (difference of only 0.20 V) to allow a better comparison. The lower difference of the gating bias suppresses the RT signal shift and positions the TE inversion domains (areas with zero thermovoltage) at approximately the same locations for both doping regimes. Interestingly, the inversion domains related to the half bow tie constrictions (regions 1 and 3) are shifted closer to the central narrowed area, instead of the inversion happening at the smallest width of the symmetric constriction as shown in previous studies [1]. This zero-shift accounts for the geometrical asymmetry influence on the thermovoltage, which, at the rectangular side of the constriction, is not drastically influenced by the further constant-distance structure edges, while, at the half bow tie side, the closer edges converge as the tip approaches the smallest width, meaning a decreased mean free path and modified thermovoltage.

The lower sample temperature achieves an enhancement of the thermovoltage range of up to 3 mV compared to the ~ 500 μV at RT due to the higher thermal gradient caused by the much hotter tip (with respect to the sample's temperature). Moreover, the polarity of the n-doped map remains the same as compared to the RT measurements amongst all the analysed area, however, the polarity of the central p-doped constriction is inversed in the LT measurements owing to the difference of gating potential. The extracted profiles also show a higher relative thermovoltage difference at areas 2 and 3 related to their geometrical disorder, similar to the RT measurements.

In summary, the thermovoltage results for both RT and LTs clearly show a modulated TE behaviour of the encapsulated graphene close to the constrictions. This phenomenon originates from the reduced mean free path of charge carriers due to the stronger influence of scattering sources. The TE-modified areas can therefore be explained as junctions between two structures with different S ($S_{graphene}$ and

$S_{constriction}$), which could lead to the fabrication of nanoscale thermocouples or other quantum energy conversion devices.

6.3.2 Hole Arrays in Tin Diselenide

Device Fabrication and Preliminary Characterisation

Due to its ultralow thermal conductivity, $SnSe_2$ has recently started to be investigated as a potential TE post-TMD. Theoretical simulations have predicted a strong dependence of the thermoelectric properties of this semiconductor material with its charge carrier concentration, achieving an increase of the *ZT* values at large temperatures [16]. The doping in this material is mostly performed through substitution with other elements, minimising the possibility of affecting the Seebeck coefficient of $SnSe_2$ with charged surface states or capacitive coupling due to its semiconductor bandgap [17]. This allows not only to study its thermoelectric properties by the standard STGM method but also to avoid the requirement of encapsulation discussed in the previous section. The electrical influence of the tip can be avoided by having a thin oxidation layer over the $SnSe_2$ and by the characteristic low frequency AC voltage applied on the probe, which together result in no current being transmitted from the tip to the layered material.

The capacitive gating influence of the probe should also be considered since it can cause a local modulation in the surface electronic transport of $SnSe_2$. However, as seen in the previous section, the related measurement difference is minor (3%) for the thin encapsulated structure. Given that the studied $SnSe_2$ flake is much thicker, one can assume that this error will be even lower as the surface-related phenomena is obscured by the thermoelectric behaviour of the entire layer. Besides the composition influence on the thermoelectric properties of the layered materials studied here, it is interesting to understand how circularly constricted areas affect the local thermovoltage and how these constrictions modulate each other if they are at different proximities. Thus, this section targets the study of hole arrays patterned into the $SnSe_2$ layer with various respective distances (pitch) and number of holes.

Bottom contacts are employed in this sample, hence, the first fabrication step is a metal deposition (see Sect. 3.4.2) following an established pattern after the 300 nm SiO_2/silicon substrate has been adequately prepared. The contacts are composed of a 50 nm gold layer on a 5 nm titanium to enhance their adhesion to the substrate. Afterwards, the $SnSe_2$ flake is mechanically exfoliated on the patterned substrate and centred over the electrodes (Fig. 6.6a). A topographical AFM scan is conducted at the edge of the flake to determine its $\approx$ 167 nm thickness. This higher thickness can also contribute to minimising the influence of the substrate's surface charge doping, given the increased amount of interlayer vdW interfaces between the bottom and top monolayers.

The different hole arrays are directly patterned via FIB milling at the centre of the $SnSe_2$ flake and close to the left electrode (Fig. 6.6b). FIB is known to disrupt the

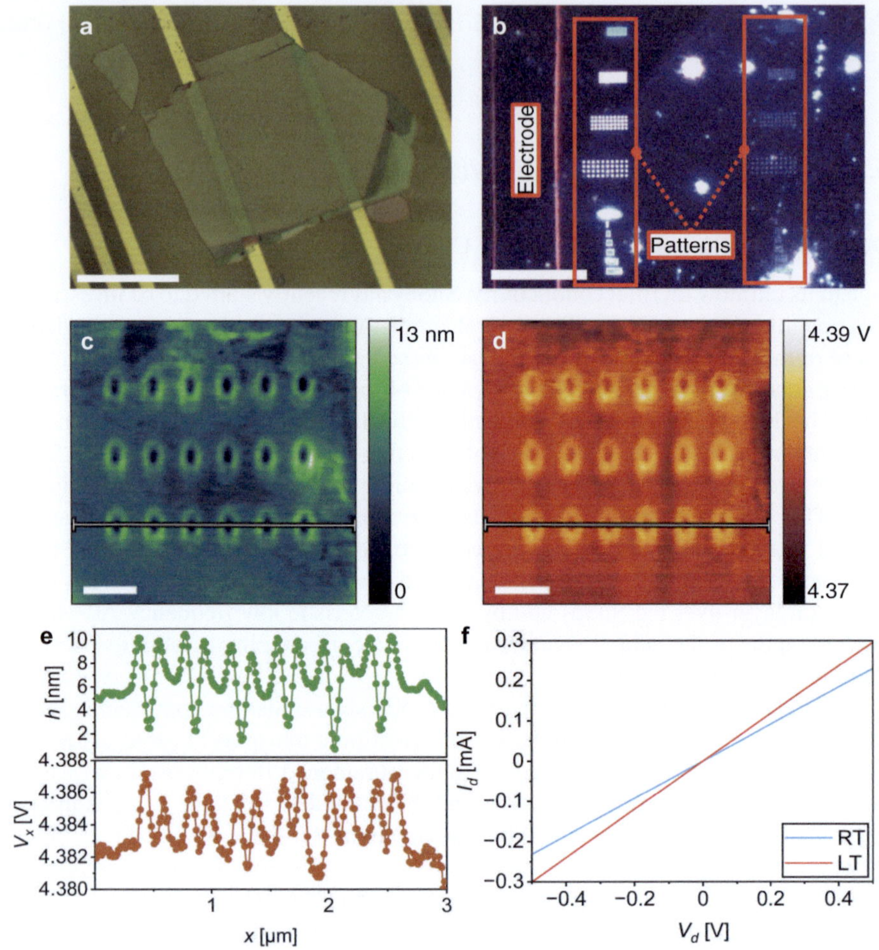

Fig. 6.6 Preliminary characterisation of the sample. **a** Optical image of the $SnSe_2$ flake on the SiO_2/silicon substrate and metallic electrodes (scale bar is 100 μm), **b** darkfield optical microscopy image of the fabricated patterns (scale bar is 20 μm), **c** topography and **d** HV SThM thermal signal maps of a 3×6 selected hole array (scale bars are 500 nm), **e** corresponding topography and thermal signal extracted profiles and **f** device current (I_d) versus voltage (V_d) response for both RT and LT regimes

chemical composition of the patterned material's surface due to its aggressiveness, hence, the ionic current of the incident gallium is reduced to 2 pA to minimise such surface damage. Nevertheless, 5–20 nm sized amorphous regions are created encircling the holes, observable at both the topographical and SThM thermal signal maps and their corresponding profiles (Fig. 6.6c–e).

The AFM topography scan of the selected 3×6 pattern shows an amorphous accumulation of sputtered material around the holes in ring shapes, increasing the topography height by ~ 5 nm with respect to the flake's reference plane. A thermal signal increase around the holes is also observed in the SThM map, denoting a decrease of the thermal conductivity once in the amorphous region due to a higher degree of boundary scattering. By analysing the SThM profile, it is possible to see that the interface between the amorphous and crystalline regions does not have a sharp transition due to a mixed-phase state, however, the majority of the areas between the holes maintain a crystalline structure, allowing their comparison with the pristine $SnSe_2$. Additionally, such amorphous regions are treated before the STGM scans with mechanical SPM sweeping, eliminating the ridges and reducing the amorphic influence on the TE properties.

Finally, the electrical characterisation of the device is performed at HV RT and LT conditions via a Keithley 2450 SourceMeter SMU unit, obtaining the current (I_d) versus voltage (V_d) curves shown in Fig. 6.6f. A symmetric (non-rectifying) linear increase is observed in both temperature cases with the current and voltage maintaining the same polarity, indicating that the device functions as a passive resistor with ohmic behaviour of approximately 2.16 and 1.68 kΩ at RT and LT, respectively. This linear behaviour demonstrates a suitable contact between the electrodes and the 2D material given its uniform electrical conduction, hence, no Schottky barriers (typical of other TMD/metal interactions like MoS_2 or WS_2 with gold or platinum) [18, 19] or trapped charges between the electrodes and the vdW structure are present in the current device. Also, the decreased resistance at low temperatures indicates a higher charge carrier mobility due to the more prominent scattering effect at elevated temperatures compared to their activation energy influence.

Room Temperature Probing of the Thermoelectric Response of the Patterned Hole Arrays

The TE characteristics of the hole arrays are measured following the procedure explained in Sect. 6.2 under HV RT conditions. Also, a similar analysis is conducted under LT conditions in A.4. To deconvolute the influence of proximal constrictions and their quantity from the local thermovoltage response, two different sets of patterns are measured with STGM.

Figure 6.7 shows both the topographical and thermovoltage response of four different 4 $\times$ 10 hole arrays with 70 nm of diameter. The holes in each of the patterns are separated from each other by distances of 400, 600, 800 and 1000 nm, respectively, creating matrices with different pitches and increasing the influential distance between the rounded constrictions. Extracting the topographical profiles from a single line of holes for each of the pitch values (green plots in Fig. 6.7c–f), it is possible to observe a decrease of the amorphous ridges encircling the holes due to the SPM sweeping, reducing from 5 to 2–3 nm.

By superimposing the topographical and thermovoltage maps (Fig. 6.7a and b, respectively), a concatenation of V_{TE} fluctuation at the constrictions is defined between the holes' rows, following the potential difference between the electrodes. This agrees with the thermovoltage gradients appearing in the constrictions of the encapsulated structure due to edge scattering. Here, two interesting phenomena are

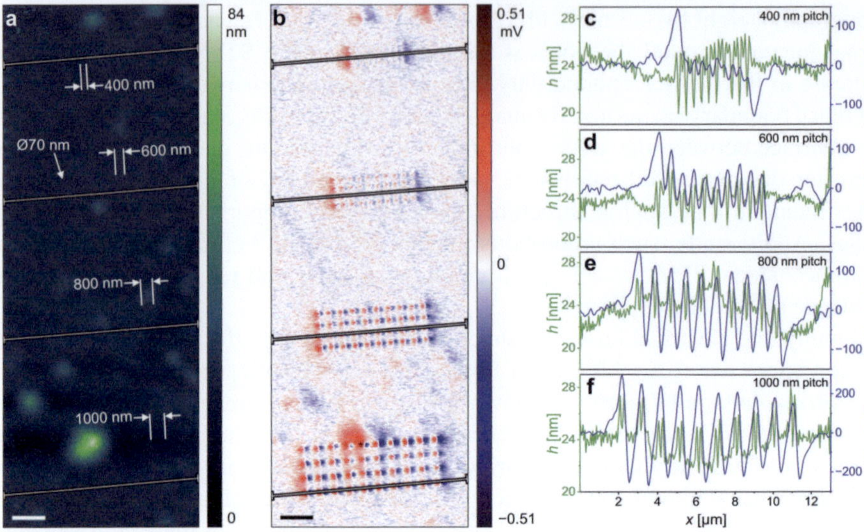

Fig. 6.7 RT HV STGM analysis of the different pitch hole arrays. **a** Topography and **b** thermovoltage maps of the patterns (scale bars are 2 μm) and **c**, **d**, **e** and **f** corresponding extracted profiles for the matrices with 400, 600, 800, 1000 nm of pitch, respectively

observed both at the thermovoltage map and at the corresponding profiles (Fig. 6.7b and blue lines in Fig. 6.7c–f, respectively).

The first one is the reduction of the V_{TE} response together with decreased pitch. A destructive interference of the V_{TE} is attributed to said effect, where, if two or more constrictions are placed nearby in the direction of the potential difference, their inverse thermovoltage polarity suppresses most of the S variation, even though edge scattering should be occurring anyway due to the presence of patterned boundaries. This can happen up to a point where, if the holes are sufficiently separated, their individual constriction TE gradient is not influenced by the surrounding TE modulation. In order to fully understand the nature of such destructive thermoelectric interference, detailed theoretical modelling is required. However, based on the previously explained principle of constrictions acting as areas with different S, a possibility would be an overlapping of the band structures at the junction compensating towards the reference S of the unpatterned region.

The second phenomenon is the presence of inversed polarity thermovoltage gradients at the short edges of the pattern rectangles, having a larger amplitude than the individual constriction thermovoltage gradients for the three patterns with shorter pitch, while at the 1000 nm pitch array, the maximum voltage of these edge peaks is similar to the constrictions' response. This could be related to the first phenomenon as an inter-hole distance limit at which the destructive interference stops occurring, thus, the hole arrays stop having a TE influence in conjunction and start behaving individually accounting principally for their S decay length. This could indicate that, due to the proximity of the holes, most of the scattering and Seebeck variation is produced at the short sides of the rectangles (the whole pattern behaving as one big rectangular constriction) with non-modulated TE behaviour inside of the array. Therefore, each of these two phenomena can be understood as additional geometrical degrees of freedom to fine-tune the local Seebeck coefficient and the thermovoltage intensity.

The second set of STGM-measured patterns corresponds to six-hole matrices having 3 rows and 1, 2, 3, 4, 6 and 8 columns, respectively, all with 70 nm of diameter. Topographical and thermovoltage maps are also extracted and shown in Fig. 6.8a and b, respectively. The lateral pitch of the patterns is kept at 400 nm, which, compared to Fig. 6.7b, resulted in a better contrast inside the rectangles, suggesting that the destructive interference not only depends on the distance between the constrictions but also on their quantities. Interestingly, the amplitude of thermovoltage at the constrictions is decreased as the number of holes increases (see Fig. 6.8c–h), in agreement with the 10-column matrices shown before. Moreover, a sloped thermovoltage transition can be seen between the short edges of the patterns when the number of holes is decreased, accounting for the proximity of the aforementioned short-side TE gradients with inversed polarity and the decay length of the Seebeck coefficient characteristic of the material. These gradients are typical of the short pitch patterns and can be observed maintaining a similar amplitude independently of the quantity of holes, only with the 3 × 1, 3 × 2 and 3 × 3 arrays showing smaller threshold values. This suggests that such gradients only occur as compensation for the destructive interference due to the nearby holes.

Other factors such as the geometrical disorder in the hole array may affect the thermovoltage local distribution in circular constrictions. A new pattern with randomly distributed 100 nm holes forming a square is fabricated on the same sample to study

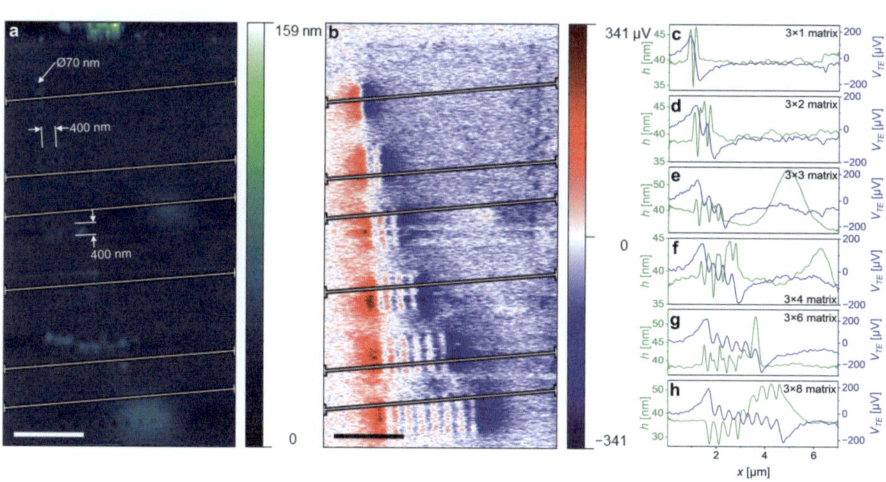

Fig. 6.8 RT STGM analysis of the arrays with different number of holes. **a** Topography and **b** thermovoltage maps of the patterns (scale bars are 2 μm) and **c**, **d**, **e**, **f**, **g** and **h** corresponding extracted profiles for the 3-row matrices with 1, 2, 3, 4, 6 and 8 columns of holes, respectively

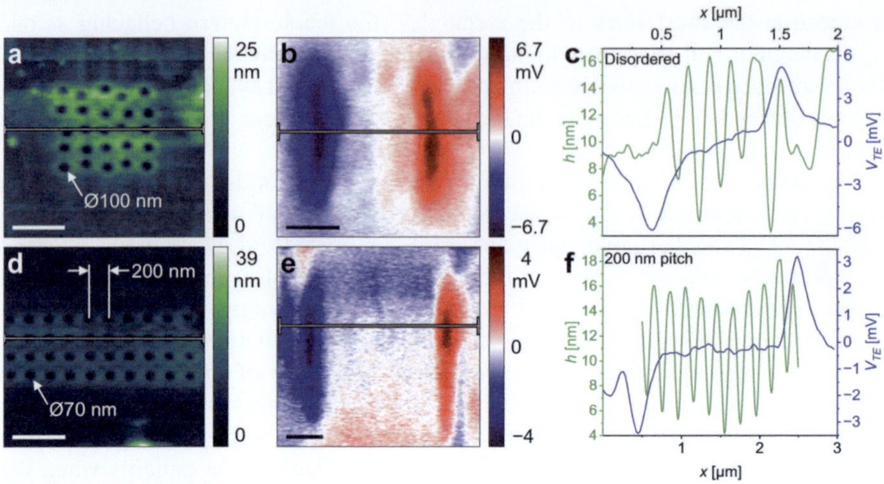

Fig. 6.9 RT STGM analysis of near intrapattern hole arrays. **a**, **d** Topography and **b**, **e** thermovoltage maps (scale bars are 500 nm) and **c**, **f** corresponding extracted profiles of the disordered and 200 nm pitch patterns, respectively

such disorder influence (Fig. 6.9a). As the diameter of the holes is increased, the amorphous ridges are higher (approximately 4–6 nm). Nevertheless, this does not seem to affect the thermoelectric response, which shows no thermovoltage constriction-related gradients inside of the square, in line with the previously discussed destructive interference of close proximity scattering interfaces. In spite of this, the edge gradients are clearly visible in both the STGM image and its extracted profile (Fig. 6.9b and c, respectively), proving that a thermovoltage modulation always appears when scattering boundaries are created on a surface, even if they are at a close distance. No influence of the disorder is then observed in the TE response. To quantify the limit at which the inter-hole distance totally eliminates the individual constriction gradients, a 4×10 pattern with holes of 70 nm in diameter and 200 nm pitch is also prepared and evaluated (Fig. 6.9d–f). As expected, no contrast is observed, thus, such 200 nm pitch value removes the response of the thermovoltage signal inside the pattern.

Thermoelectric Finite Element Analysis of the Hole Arrays

The thermoelectric response of the arrays can also be simulated via FEA using COMSOL Multiphysics. A realistic model of the sample is created by stacking the metallic electrodes and the $SnSe_2$ layer (with a similar thickness to the real flake) on top of the SiO_2/silicon substrate (Fig. 6.10a). The holes are drawn at the centre of the flake, however, each simulation contains only the specific pattern that is studied instead of all the different types of hole matrices present in the real sample. This is allowed since the patterned rectangles do not have any observable influence on each other in the STGM studies. By performing such type of modelling, the computation efficiency is greatly enhanced, and multiple compilations can be simultaneously performed, making the whole process more time efficient. The Seebeck coefficient depends on the position of the heat source with respect to the constrictions, hence, it is necessary to define S as a custom function (becoming zero from the void boundary) assuming an exponential change of the Seebeck coefficient at the vicinity of the void. For circular constrictions in arrays with different pitch and constriction quantities, this will result in:

$$S(x, y) = S_0 \prod_{i=1}^{N_h} 1 - \min\left(1, e^{-\frac{\sqrt{\left(x - x_{h,i}\right)^2 + \left(y - y_{h,i}\right)^2} - a_h}{l_S}}\right) \tag{6.3}$$

where x and y are the tip's horizontal and vertical positions, S_O is the $SnSe_2$ layer's reference Seebeck coefficient N_h is the total number of holes, i is the incremental factor, $x_{h,i}$ and $y_{h,i}$ are the horizontal and vertical positions of each hole, a_h is the hole's radius and l_S is de decay length of the Seebeck coefficient. By applying this function to the model layers in COMSOL, it is possible to obtain maps of the S behaviour of the patterned layer as shown in Fig. 6.10b for a 3 × 3 pattern of 70 nm diameter holes separated by a distance of 400 nm. As predicted, the Seebeck is null inside of the holes since no physical material is present. However, the Seebeck variation is not constant, following a gradient related to the thermovoltage that is quantified by fitting the value of the decay length with the previously explained STGM experimental results.

Thermovoltage maps can also be obtained from the same 3 × 3 structure by probing the received voltage on one of the electrodes while the other is grounded and the probe acts as a heat source scanning the 2D region (Fig. 6.10c). FEA simulations allow to compare the same phenomena observed in the experimental data and the simulated results, such as the low contrast change of the constrictions due to the hole proximity and the side thermovoltage gradients retaining the largest amplitude.

To further compare with the experimental data, one-dimensional profiles are simulated along the different sets of patterns emulating the real sample, i.e., with different pitch and number of holes, in Fig. 6.10d–n. The same destructive interference behaviour is observed with the amplitude of the inner peaks decreasing together with the pitch and the number of concatenated constrictions. Also, the decrease of the amplitude of the side gradients is more noticeable with the FEA simulations for the case of the pitch increment, validating the reduced effect between the individual constrictions as they become more separated from each other. For the variation of hole quantity, the first noticeable behaviour is the steep slope on the inside of the rectangles while the polarity of the thermovoltage changes, also occurring in Fig. 6.8. An interesting phenomenon comparable to the experimental measurements is the constant peak height of the side gradients as the number of constrictions varies. This phenomenon is reproduced for all the patterns except the arrays of one- and two-hole columns, where a smaller amplitude is shown, perhaps deconvoluting the influence of nearby constrictions in the thermovoltage response.

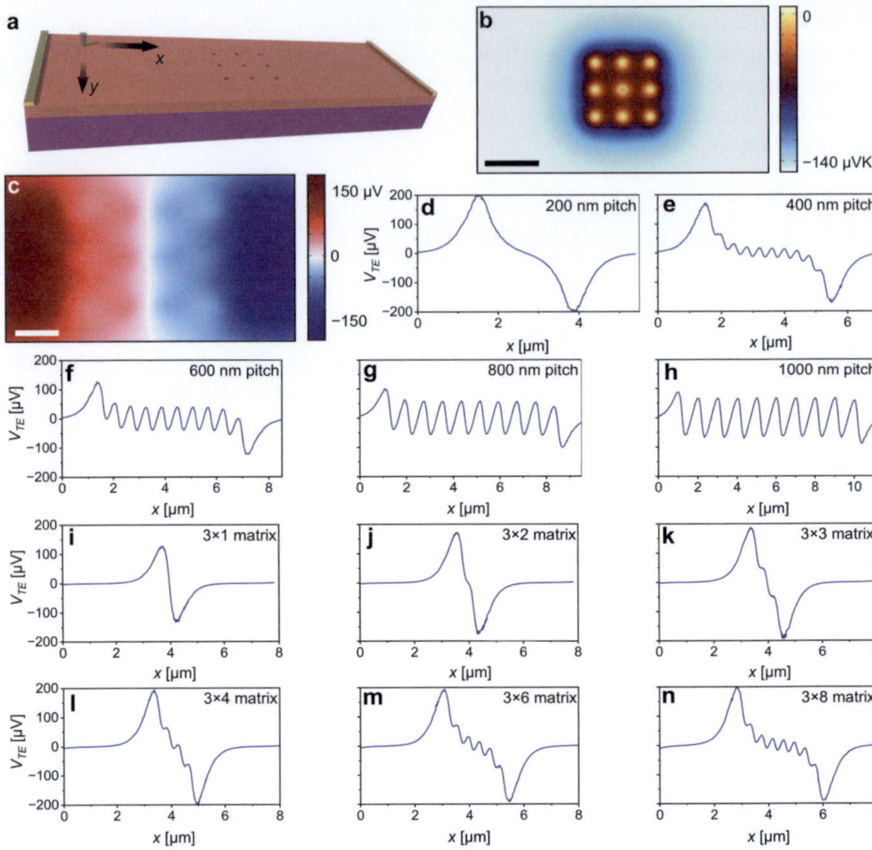

Fig. 6.10 RT FEA thermoelectric modelling of the hole arrays. **a** Model used for the FEA simulations with the two directions of the tip's movement, **b** Seebeck coefficient and **c** thermovoltage maps of the selected 3 × 3 pattern (scale bars are 800 and 200 nm, respectively), **d**, **e**, **f**, **g** and **h** single-row thermovoltage profiles extracted from the 200, 400, 600, 800 and 1000 nm of hole pitch matrices, respectively, and **i**, **j**, **k**, **l**, **m** and **n** single-row thermovoltage profiles extracted from the 3 × 1, 3 × 2, 3 × 3, 3 × 4, 3 × 6 and 3 × 8 matrices, respectively

6.4 Thermoelectric Thickness Boundaries in Graphite

The role of boundary scattering in nanoconstrictions has been extensively studied in the previous sections, however, there was no thickness variation in the measured areas. Topographical variances are interfaces that can also be considered influential sources to both phononic and electronic transport, as seen in Chap. 5. Thus, regions with different thicknesses can be understood as structures with varied thermoelectric properties. To quantify such influence, a graphite flake containing an area of convergence between three different thicknesses is mechanically exfoliated on a SiO_2/silicon substrate after the supporting layer has been properly activated and

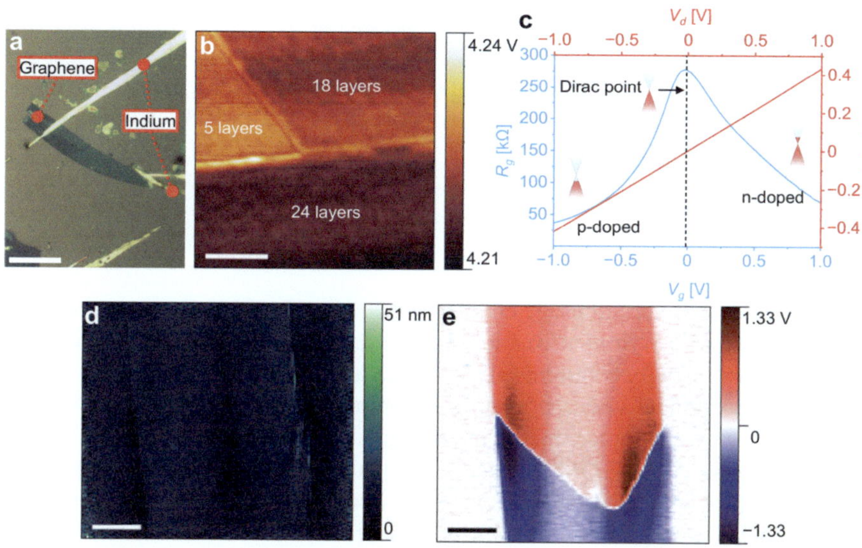

Fig. 6.11 Preliminary and STGM analysis of the graphite device. **a** Optical image of the device showing the graphite flake and the indium electrodes (scale bar is 40 μm), **b** SThM thermal signal map centred at the convergent point of the layers with different thicknesses (scale bar is 500 nm), **c** device resistance (R_d) versus gate voltage (V_g) and device current (I_d) versus device voltage (V_d) plots showing the different doping regions, **d** topography and **e** thermovoltage maps cantered at the convergence point of the graphite flake (scale bars are 5 μm)

cleaned. Afterwards, the sample is annealed inside a vacuum oven at 160 °C for 1 h to remove any remaining contaminants. To electrically bias the graphite flake, two indium contacts are deposited following the procedure explained in Sect. 3.4.2 (Fig. 6.11a) and wire bonded using silver paste. Such arrangement allows to contact arbitrary areas of 2D materials without the risk of contamination from EBL resist.

The first performed characterisation is a magnified HV SThM map, shown in Fig. 6.11b. In this scan, the convergence point is centred, and a clear variation of the thermal properties of different-thickness graphite can be observed, with an inverse relation between the thermal conductivity and the thickness due to the higher proximity of the tip to the insulator SiO_2 layer at lower topographical values. Secondly, the electrical properties of the graphite device are quantified by an I_d versus V_d curve and a R_d versus V_g plot (Fig. 6.11c), showing an ohmic linear behaviour similar to the $SnSe_2$ device and proving the feasibility of the indium connections with the graphite layer. Also, the CNP point is located almost at zero gate voltage, indicating a negligible native doping level in the sample. Lastly, the STGM topography and thermovoltage maps are measured under HV and RT conditions over an extensive area covering all the different thicknesses and the substrate, which is taken as a reference zero signal given its non-conducting behaviour. In spite of having a bigger thickness difference, there is not a clear contrast change between the 5- and 18-layers' region (parallel to the potential difference direction), only an observable gradient of thermovoltage. However, perpendicularly to the electrodes' direction and following the shape of the flake, an inverse polarisation is observed as the thickness changes to 24 layers, indicating a higher contribution of that interface as a greater scattering source, possibly due to a crack which is not observable with AFM topography.

6.5 Summary

Local thermoelectric variations are characterised for the first time in an encapsulated graphene device with asymmetric nanoconstrictions and in $SnSe_2$ circular hole arrays via STGM. Such patterned systems allow to measure the variation of thermovoltage and Seebeck coefficient with the material's composition, gate doping, temperature of the sample, constriction geometry and proximal void influence.

As stated by previous STGM studies, nanoconstrictions create a spatial-dependent thermovoltage inversed polarity region due to scattering boundaries and modified electronic mean free path. Here, this phenomenon is confirmed by the present measurements and represents a novel path for local modification of TE behaviour. It is also proven that encapsulation of graphene with electrically insulating materials is required to control the parasitic doping in the vdW layer from the substrate or the tip and preserve its intrinsic electronic properties, leading to enhanced TE response.

Backgate doping has been employed as an additional tool to control and modulate not only the sign of the device's thermovoltage but also its amplitude and spatial distribution, with a direct effect on the charge carrier behaviour of graphene. The effect of an increased temperature gradient between the hot spot and the rest of the studied structure is measured in both the encapsulated and hole-array devices, where a clear enhancement of the thermovoltage range is observed as the sample is cooled. For the hole matrices device, an influential behaviour of nearby constrictions results in destructive interferences of the thermovoltage response, which could end in the total suppression of the thermovoltage signal if the constrictions are closer than 200 nm. Moreover, side thermoelectric gradients related to the decay length of the Seebeck coefficient are observed in such matrices, and their nature could be linked to the compensation for the destructive interference between the constrictions. These two latter phenomena are validated by FEA simulations of the structure, showing a good correlation of the thermoelectric response. Finally, a graphite device has been measured using arbitrarily positioned indium electrodes, allowing to quantify the influence of thickness in the thermovoltage signal, finding that its role is not prominent in such systems compared to other scattering sources if the thickness boundary is parallel to the observed potential difference direction.

STGM has proven to be a unique local probe microscopy technique that can be employed to extract the thermovoltage signal of two-dimensional devices with distinct nanoscale geometries non-invasively and with a nanoscale spatial resolution. The required setup and calibration of the system is relatively straightforward compared to other SPM methodologies and provides direct and reliable access to local thermoelectric properties.

In general, this chapter presents multiple ways of precisely controlling the Seebeck coefficient in 2D materials, which represent a fundamental advance for local TE modulation studies at the nanoscale. This could not only evolve the functioning of established technologies such as temperature sensors, thermophotovoltaic detectors or heat management applications but also open the door for novel nanoscale systems like single metal thermocouples or RT high performance thermoelectric structures.

References

1. Harzheim, A., et al. (2018). Geometrically enhanced thermoelectric effects in graphene nanoconstrictions. *Nano Letters, 18*, 7719–7725.
2. Levander, A. X., et al. (2011). Effects of point defects on thermal and thermoelectric properties of InN. *Applied Physics Letters, 98*, 012108.
3. Xia, F., et al. (2009). Photocurrent imaging and efficient photon detection in a graphene transistor. *Nano Letters, 9*, 1039–1044.
4. Harzheim, A., Evangeli, C., Kolosov, O. V., & Gehring, P. (2020). Direct mapping of local Seebeck coefficient in 2D material nanostructures via scanning thermal gate microscopy. *2D Materials, 7*, 041004.
5. Nistor, R. A., Kuroda, M. A., Maarouf, A. A., & Martyna, G. J. (2012). Doping of adsorbed graphene from defects and impurities in SiO_2 substrates. *Physical Review B, 86*, 041409.
6. Martin, J., et al. (2008). Observation of electron–hole puddles in graphene using a scanning single-electron transistor. *Nature Physics, 4*, 144–148.
7. Duan, J., et al. (2016). High thermoelectricpower factor in graphene/hBN devices. *Proceedings of the National Academy of Sciences of the USA, 113*, 14272–14276.
8. Schaumburg, F., et al. (2023). Enhanced intensity of Raman signals from hexagonal boron nitride films. *Applied Physics Letters, 123*, 073101.
9. Wang, H., Wang, Y., Cao, X., Feng, M., & Lan, G. (2009). Vibrational properties of graphene and graphene layers. *Journal of Raman Spectroscopy, 40*, 1791–1796.
10. Karak, S., et al. (2021). Hexagonal boron nitride–graphene heterostructures with enhanced interfacial thermal conductance for thermal management applications. *ACS Applied Nano Materials, 4*, 1951–1958.
11. Zhu, W., Zheng, G., Cao, S., & He, H. (2018). Thermal conductivity of amorphous SiO_2 thin film: A molecular dynamics study. *Science and Reports, 8*, 10537.
12. Langer, G., Hartmann, J., & Reichling, M. (1997). Thermal conductivity of thin metallic films measured by photothermal profile analysis. *Review of Scientific Instruments, 68*, 1510–1513.
13. Sun, W., Liu, H., Gong, W., Peng, L.-M., & Xu, S.-Y. (2011). Unexpected size effect in the thermopower of thin-film stripes. *Journal of Applied Physics, 110*, 083709.
14. Wang, L., et al. (2013). One-dimensional electrical contact to a two-dimensional material. *Science, 342*, 614–617.
15. Goldsmid, H. J. (2010). *Introduction to thermoelectricity* (1st ed.). Springer-Verlag.
16. Pham, A.-T., et al. (2020). High-quality $SnSe_2$ single crystals: Electronic and thermoelectric properties. *ACS Applied Energy Materials, 3*, 10787–10792.
17. Huang, Y., et al. (2016). First-principles study on doping of $SnSe_2$ monolayers. *ChemPhysChem, 17*, 375–379.
18. Pollmann, E., et al. (2021). Large-area, two-dimensional MoS_2 exfoliated on gold: Direct experimental access to the metal–semiconductor interface. *ACS Omega, 6*, 15929–15939.
19. Tang, H., et al. (2019). Schottky contact in monolayer WS_2 field-effect transistors. *Advanced Theory and Simulations, 2*, 1900001.

Chapter 7
Thermal, Electrical and Mechanical Cross-Interaction in Nanostructures

Abstract In this chapter, the coupled thermal, electrical and mechanical physical phenomena within complex nanostructures is investigated, focusing on two distinct systems: twisted double bilayer graphene (TDBG) and nanotube-based buckypapers. The first part of the chapter explores the electromechanical behaviour of TDBG, employing advanced scanning probe microscopy techniques to characterise local flexoelectricity and nanomechanical responses for the first time. Emphasis is placed on identifying Moiré superlattices capable of piezoelectric behaviour, detecting electromechanical phase delays and probing stiffness variations related to stacking order. These findings are supported by force field relaxation simulations, which provide insight into the local deformation and strain distribution around domain walls. The second part targets the thermomechanical analysis of one-dimensional carbon and boron nitride nanotube buckypapers, focusing on samples with a coaxial nanotube architecture. This section includes the synthesis protocols and morphological characterisation of the networks, followed by a novel application of piercing scanning thermal microscopy (pSThM) to evaluate their thermal transport behaviour under mechanical load. The results reveal enhanced thermal performance and mechanical adaptability, highlighting the potential of such soft nanomaterials in multifunctional applications. Together, the studies presented in this chapter offer new insight into cross-interactions in nanoscale systems and the role of SPM in exploring complex multiphysical behaviours.

7.1 Introduction

Over the previous chapters, the thermal transport characteristics and the interplay between electronic and heat transport in vdW materials have been extensively studied. However, another crucial aspect of these structures is their mechanical behaviour and the degree of influence on the other two mentioned properties, especially, when it comes to the integration of such materials in real-world developments. The anisotropic nature of 2D materials is one of the most important sources of their unique set of properties, given the different bonding configurations between the in-plane

S. Gonzalez-Munoz, *Interplay of Nanoscale Physical Properties in 2D Materials and Heterostructures*, Springer Theses, https://doi.org/10.1007/978-3-032-14271-9_7

and cross-plane directions, hence, also affecting their mechanical response. Such binding forces are consequently responsible for the changes in Young's modulus, tensile strength, hardness, durability, etc., but, given that they arise between elements capable of heat and charge transport, strong interlinks are generated between the material's electronic and thermal features with its mechanical behaviour, ultimately entering in the electromechanical and thermomechanical fields.

Novel approaches are being developed to enhance or modulate the electro- and thermomechanical capabilities in vdW systems while comprehending their intrinsic nature. In this sense, the field of twistronics, which has gained increasing scientific interest in recent years, can play a significant role. By stacking two or more layers of 2D materials and rotating them a certain relative angle, the electrical, magnetic and optical characteristics of the resulting heterostructure are precisely controlled [1–3]. However, the nanomechanical and electromechanical response on said systems has hardly been investigated [4–6]. Their resulting Moiré superlattice configuration has an intrinsic strain accumulation at the domain boundaries due to the tendency to return to a relaxed state, which causes a significant impact on their mechanical properties and can affect the electronic transport by the frustration of flat bands' formation [7]. Such superlattice strain causes domains with commensurate and incommensurate transitions between the composing layers, raising the requirement of performing further investigation on additional structures other than graphene/h-BN [8].

Other interesting approaches for the leverage of multiphysical performance in nanoscale materials include their configuration as buckypapers. Buckypapers (BP) are 1D nanotube-based tangled networks that form a macroscale paper-like porous thin film with vdW binding between the single (SW) or multiwalled (MW) nanotubes. Although the term was originally monopolised by carbon nanotube-based (CNT) thin films [9], currently, the composition of BPs includes other elements like boron nitride nanotubes (BNNT) [10, 11]. Such configuration allows a controlled enhancement of, e.g., the thermal conductivity along the nanotubes' axes, which can also be extrapolated at the macroscopic scale by the interconnections between the nanotubes [12]. Also, the mechanical properties of BPs, i.e., high flexibility and stability as well as lightweight and porous structure make them easier to manipulate, homogeneously combine with other elements forming nanocomposites and adapt to non-uniform shaped frames [13].

BNNTs have been less synthesised and studied than CNTs, however, it is known that the former presents an insulating electronic behaviour compared to the chirality-dependent semiconducting or metallic charge transport in CNTs. Their thermomechanical characteristics also differ, given that BNNT owns a superb elastic modulus as well as thermal stability and can resist oxidation up to elevated temperatures [14, 15]. A synergetic anisotropy-controlled combination of both CNT and BNNT as coaxial heteronanotubes with a graphene core and an h-BN shell (CNT@BNNT) can result in the advantageous thermal and mechanical properties of BNNT merging with the superb electronic transport in CNT (along the nanotube direction) and dielectric cross-plane BNNT shielding. This positions both BNNT and CNT@BNNT as promising candidates for heat-dissipating applications.

Therefore, the main idea encompassing this chapter is to probe the interplay of the electrical and thermal phenomena with the mechanical behaviour in two types of nanoscale structures. The first part of the chapter focuses on the electro- and nanomechanical quantitative investigation of twisted double bilayer graphene (TDBG) via different SPM techniques, with an additional force field relaxation modelling for nanomechanical comparison. Such twisted homostructure contains commensurate stacked Moiré domains separated by single and double domain walls (SDW and DDW, respectively), inducing regions with locally varied cross-plane Young's moduli [16]. Also, it has been recently investigated as a material with flexoelectric polarised domains [17]. Thus, this system offers an extensive platform for conducting localised nanoelectromechanical investigations. The second section of this chapter is dedicated to a novel synthesis methodology of three BP types (MWCNT, MWCNT@BNNT and MWBNNT) based on atmospheric pressure CVD, vacuum filtration and thermal oxidation and their respective thermomechanical study via piercing scanning thermal microscopy (pSThM) reported here for the first time. The fabrication procedure is characterised by being economic and feasible for scaled production. Also, it involves a singular source ammonia borane precursor, enhancing the mechanical stability and eliminating hazards from other employed precursors. pSThM provides nanoscale resolution thermal response, allowing to extract the heat conductivities with respect to the softness of the BP and achieving an enhancement of this parameter for MWCNT@BNNT.

In the first part of the chapter, the SPM measurements and data analysis were performed by myself in the Lancaster University Physics Department with the acknowledged collaboration of Alessandra Canetta, Dr. Jean Spièce and Prof. Pascal Gehring from the Catholic University of Louvain, while the force field simulations were performed by Dr. Viet-Hung Nguyen and Prof. Jean-Christophe Charlier, also from the Catholic University of Louvain. For the second section, the sample fabrication and preliminary characterisation were performed by Dr. Ruth S. Jones, Dr. Barbara M. Maciejewska and Prof. Nicole Grobert at the University of Oxford, whereas the pSThM and its resulting data analysis were performed by myself in the Lancaster University Physics Department. The results exposed in this chapter have been published in Canetta et al. [37] and Jones et al. [38] (see "List of Publications").

7.2 Electromechanics of Twisted Double Bilayer Graphene

7.2.1 Piezoelectric Behaviour and Stacking Order

Sample Structure and Piezoresponse Force Microscopy Custom Setup

SPM mechanical lithography is employed to cut two parts of 2LG, which are picked up and twisted at a relative angle of $\sim 1^\circ$ following the dry transfer stamping procedure (see Sect. 3.3), with an additional layer of poly(carbonate) (PC) attached to the PDMS to increase the addition with the vdW materials. The TDBG stack is deposited on a pre-exfoliated h-BN flake supported by a SiO_2/silicon standard substrate. This deposition is performed at RT to avoid the melting of the PC layer, leaving polymeric contamination on the surface and positioning the TDBG at an elevated angle with respect to the h-BN to decrease its sub-Moiré contribution influence.

Each of the two stacked 2LG films is arranged with an AB Bernal configuration, and their relative angle is close to zero to increase the periodicity of the generated Moiré pattern (with a lattice constant of β_M). With the superimposition of the two Bernal-stacked bilayers, three new states are created (Fig. 7.1a), namely ABAB (Bernal), ABCA (rhombohedral) and ABBC. The local BB stack between the second and third layers of the ABBC configuration is energetically unfavourable, resulting in the ABBC stacking to be the most unstable and promoting the spatial expansion of the other two configurations [18]. Thus, ABAB and ABCA discrete domains create commensurate regions separated by transitional areas called saddle point (SP) boundaries or domain walls [17]. Comparing the energetic states of the ABAB and the ABCA domains, the former presents a higher stability, hence, the energy imbalance between them forces the ABAB and ABCA regions to adapt to convex and concave forms, respectively [17]. Also, this imbalance promotes the merging of two SP domain walls into a double boundary (DSP) separation, with the interlayer strain caused by the twist angle facilitating this convergent phenomenon [16]. Thus, the resulting structure can be understood as SP transitions separating commensurate domains of different stacking order (ABAB and ABCA) and DSP boundaries splitting commensurate domains of the same stacking type (ABAB) [16].

To probe such stacking domains and separation walls as well as the layers' flexoelectric response, a Bruker MultiMode® Nanoscope® IV AFM is used and configured for piezoresponse force microscopy (PFM) with a custom electronic setup (Fig. 7.1b) and ambient conditions. In PFM, an AC bias ranging from 1 to 3 V is supplied by a Rigol DG1062Z 60 MHz Function/Arbitrary Waveform Generator (GEN) between the conductive tip (NANOSENSORS™ PtSi-CONT) and the sample to induce an electric field in the cross-plane direction. The frequency is modulated to adjust to the contact resonance frequency between the tip and the sample (~ 80 kHz). The

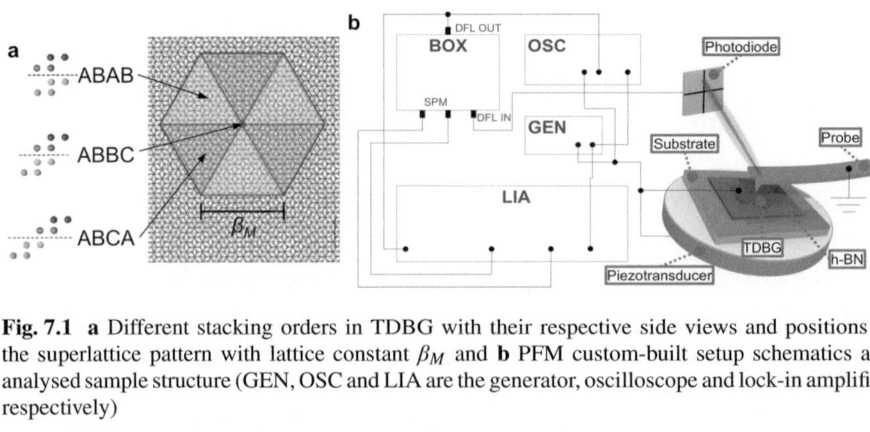

Fig. 7.1 **a** Different stacking orders in TDBG with their respective side views and positions in the superlattice pattern with lattice constant β_M and **b** PFM custom-built setup schematics and analysed sample structure (GEN, OSC and LIA are the generator, oscilloscope and lock-in amplifier, respectively)

electric field causes a periodic strain in the piezoelectric layers of the sample, which is tracked by the vertical deflection of the cantilever at the stipulated resonance frequency (see Sect. 2.4.2 for a detailed explanation). The resulting signal is sent through a conversion box (BOX) to the LIA (Stanford Research Systems SR830 DSP), which deconvolutes the electromechanical response into amplitude and phase and sends both contributions again through the BOX to the SPM controller.

Analysis of the Flexoelectric Domains

As previously reported in the literature [19], a coupling between the piezoelectric cross-plane electric field and the interlayer strain gradients induced by the twist angle allows the visualisation of PFM contrast. This same principle is applied here to image the discrete stacking domains of the TDBG sample.

The obtained amplitude and phase PFM maps of a flat surface in the TDBG structure are shown in Fig. 7.2a and b, respectively, where elongated triangular domains are distinguished. However, the maximum contrast difference is not defined at the domains as such, instead, the domain interfaces are the ones that show a higher signal variation. Such interfaces correspond to the previously mentioned SP and DSP boundaries, thus, attributing the concave and convex separated domains as ABCA and ABAB stacking configurations, respectively [20]. The specific domain

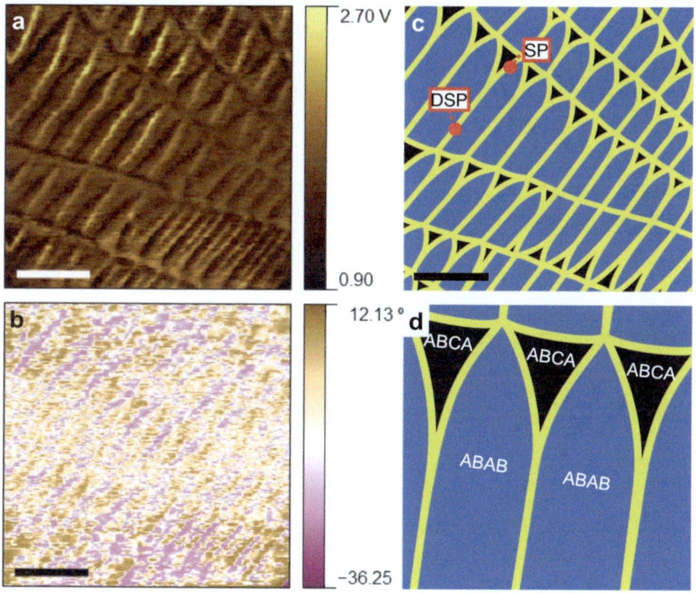

Fig. 7.2 PFM imaging of the reconstructed lattice and stacked domains. **a** Amplitude and **b** phase PFM maps of an area in TDBG showing different stacking domains (scale bars are 200 nm), **c** scheme of the different domain types extracted from **a** showing the two types of superlattice boundaries (scale bar is 200 nm) and **d** magnified view of a region in **c** showing the multiple stacking orders present in the sample (blue and black) being delimited by the domain walls (yellow)

type distribution is depicted in Fig. 7.2c and d, where blue and black regions indicate ABAB and ABCA domains, respectively, whereas the yellow lines represent their boundaries. By a quantitative analysis of such figures, the reconstructed lattice constant (β_M) can be obtained and used in the following equation to extract the twist angle [21]:

$$\beta_M = \frac{\beta_g}{2}\csc\left(\frac{\theta}{2}\right) \tag{7.1}$$

where β_g is the lattice constant of graphene and θ is the calculated interlayer twist angle. Thus, the effective inter-bilayer twist angle is quantified as $0.11 \pm 0.04°$. The stretched nature of the triangular geometries in the superlattice is attributed to the combined influence of non-uniform strain originating from both the atomic relaxation given the minimal twist angle and external geometrical factors such as proxime wrinkles or material edges [16]. Also, the PFM amplitude map shows the convergences of the two SP domain walls surrounding ABCA domains into DSP boundaries separating the stretched ABAB adjacent domains, agreeing with the literature [16]. However, the difference in contrast of the domain walls (corresponding to the edges of the PFM amplitude scale) represents an electromechanical polarisation based on a local flexoelectric behaviour of TDBG subjected to the electric field and the strain gradient, also supported by the phase map. In piezoelectricity, only a homogeneous deformation can cause an electrical response, while in the current sample, the strain is non-uniform due to the different stacking orders, atomic relaxations and domain walls, hence, breaking the original centrosymmetry of the structure causes a defined polarity [17, 19]. Therefore, by controlling the twist angle between the two 2LG films, not only the amount of flexoelectric deformation can be modulated but also its local distribution, resulting in nanoscale electromechanical effects applicable to, for instance, ultra-precise sensors or actuators [22].

7.2.2 Electromechanical Phase Delay Detection

Electric Heterodyne Force Microscopy Configuration

Classical heterodyne force microscopy (HFM) [23, 24] allows to explore the dynamic mechanical properties of the sample, such as its viscoelasticity or adhesive response (see Sect. 2.4.2), but it can also be modified to probe the dynamic electromechanical behaviour of piezoelectric materials. This is done by adding an electrical connection to the sample (similarly to PFM) to the HFM setup so that the transversal electric field creates a deformation of the studied layers tracked by the cantilever's deflection. In order words, PFM and HFM combined generate an SPM method named electrical heterodyne force microscopy (E-HFM), capable of non-linear detection of ultrasonic vibration and local piezoelectricity, reported here for the first time.

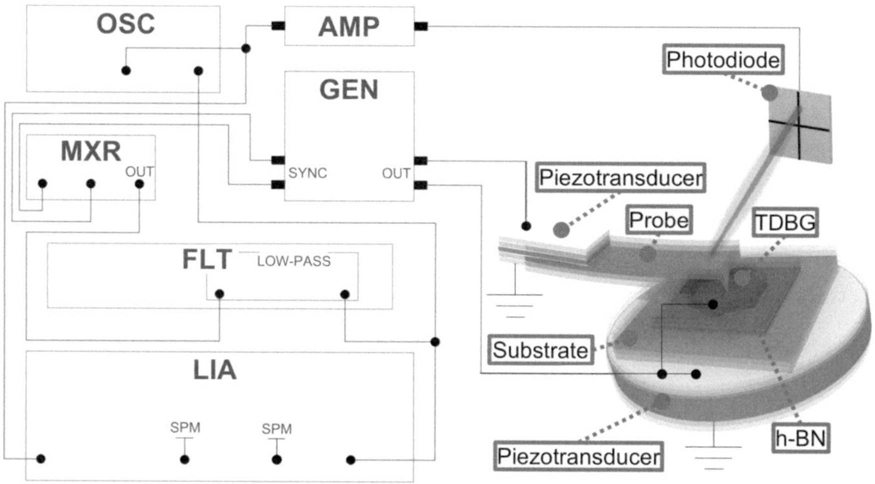

Fig. 7.3 E-HFM custom-built setup schematics and analysed sample structure (GEN, AMP, OSC, LIA, MXR and FLT are the generator, amplifier, oscilloscope, lock-in amplifier, mixer and filter, respectively)

The E-HFM setup is shown in Fig. 7.3 (Bruker MultiMode® Nanoscope® IV AFM under ambient conditions) and consists of the separate ultrasonic excitation of both the tip ($\omega_1 = 4$ MHz) and the sample ($\omega_2 = 4.30$ MHz) via their respectively attached piezotransducers (bonded with phenyl salicylate, also known as salol). Simultaneously, the AC bias created to excite the sample is also used to induce the transversal electric field as in PFM. The two sinusoidal excitation signals are applied with a Rigol DG1062Z via its dual channel configuration, which also supplies the respective synchronous transistor–transistor logic (TTL) reference signals with the same frequencies and phases. Due to the distinct vibration frequencies of the sample compared to the probe, the maximum tip/surface distance is modulated at their difference ($|\omega_1 - \omega_2|$), known as the beat frequency. A custom-built electronic mixer (MXR) is used to build such difference frequency from the synchronous signals, creating a reference beat frequency which is then filtered through the Stanford Research Systems SR650 FLT and supplied to the reference input of the Stanford Research Systems SR830 DSP LIA. When the total amplitude is able to cover the entire non-linear range of the tip/sample contact force, an ultrasonic force contribution dependent on such amplitude (comparable to UFM in Fig. 7.5b and c) and a force caused by the piezoelectric deformation are added to such interaction strength influencing the cantilever by creating an additional displacement of the tip. This is controlled by monitoring the cantilever's deflection with the LIA after being preamplified by a Stanford Research Systems SR555 AMP, which can be directly related to the additional force by dividing the latter by the cantilever's stiffness (knowing that the beat frequency is kept below the resonance frequency of the probe and any cut-off detection frequencies of the system) [24]. Hence, the beat frequency changes at a particular time resolution with respect to the modulated reference value.

Similarly to PFM, E-HFM monitors the amplitude and phase shift of the varied beat frequency. The amplitude maps provide nanoscale variations of the sample's elastic and piezoelectric behaviour, representing both the absolute electro- and nanomechanical properties of the scanned materials. Yet of greater significance are the phase shift maps, capable of high resolution temporal sensitivity (at a fraction of the ultrasonic time scale). These maps provide access not only to time-dependant (dynamic) viscoelastic or adhesive response but also to the electronic dynamics related to deformation and relaxation [24].

Heterodyne Subsurface Response and Relaxation Regimes

The resulting E-HFM maps of TDBG are shown in Fig. 7.4. The topographical scan (Fig. 7.4a) is able to probe maximal features of ~ 460 pm, indicating the elevated flatness of the layers. One can observe a high similarity between the amplitude images of both PFM and E-HFM scans (Figs. 7.2a and 7.4b, respectively) due to the combined influence of the electro- and nanomechanical phenomena. However, the nanomechanical contrast (see Sect. 7.2.3) is much lower than the electromechanical one, indicating a higher effect of the flexoelectric properties over the changes in elastic behaviour. Nevertheless, a change in the contrast between the ABAB and ABCA domains (not observed in PFM) is distinguished here. One could attribute this to the different stability of the domains, however, the behaviour should be the opposite since the ABAB domains have higher stability (corresponding to a brighter contrast) than the ABCA neighbours. Also, the UFM measurements described in the next section show practically no variation in the pure elastic response of both regions. Thus, this phenomenon could indicate other external influences such as electrostatic interactions between the sample and cantilever (to which E-HFM is much more sensitive than PFM and UFM) [24]. Further measurements with other dynamic SPM techniques or simulations would be necessary to fully comprehend the origin of such contrast. The phase shift image (Fig. 7.4c) can provide more reliable information about the elastic and electromechanical properties, with the extra insight into the time-dependent behaviour. By looking at the scale range, the contrast shift is approximately $\Delta\theta = 12.40°$, hence, applying the following equation, the time delay (Δt) corresponding to such phase shift can be calculated:

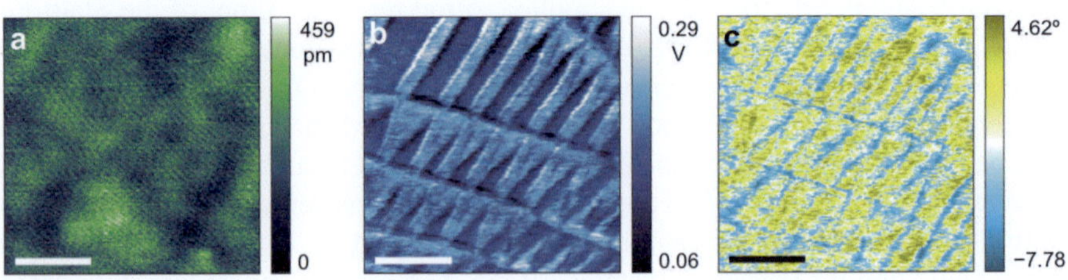

Fig. 7.4 E-HFM response of the TDBG superlattice. **a** Topography, **b** E-HFM amplitude and **c** E-HFM phase maps of the analysed region. Scale bars are 200 nm

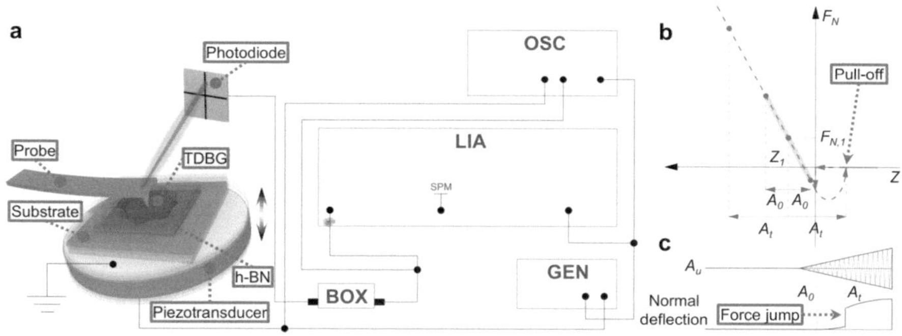

Fig. 7.5 **a** UFM setup schematics and analysed sample structure (GEN, OSC and LIA are the generator, oscilloscope and lock-in amplifier, respectively), **b** normal force vs. tip/sample separation distance curve and **c** typical input (ultrasonic excitation A_u) and output (normal deflection) UFM signals

$$\Delta t = \frac{\Delta\theta}{360\omega_2} \tag{7.2}$$

resulting in a delay of 8.01 ns between the yellow versus blue areas. To compute the influence of the electromechanical and the nanomechanical contributions in such time delay, one can also calculate the propagation of the mechanical wave in TDBG, knowing that the thickness is $\approx$ 13.40 Å and its transversal sound velocity is 12.90×10^3 ms^{-1} [25, 26]. Thus, the mechanical delay is 0.21 ps, four orders of magnitude lower than the total time delay. It is therefore possible to argue that the majority of the delay is caused by the electromechanical relaxation phenomena, in agreement with the amplitude image. This ns time scale also gives an insight into how fast the charges move through the layers for each of the domain and domain wall types. The contrast is mainly observed between the domains and their boundaries, indicative of their different relaxation times after the electronic excitation. As the domain walls show a larger phase shift, their relaxation time increases, provided their incommensurate state.

7.2.3 *Probing the Nanomechanical Response*

Ultrasonic Force Microscopy Setup

After the electromechanical analysis of the structure, its nanomechanical response must also be measured with the ultimate goal of defining the local Young's modulus variations expected from domains with different energies, as well as their corresponding stacking orders to link them to their flexoelectric behaviour.

To perform such tasks, an ultrasonic force microscope (UFM) with a custom setup is employed (Fig. 7.5a) on the same ambient condition AFM used for the electromechanical measurements. This technique uses an AFM tip (BudgetSensors® ContAl-G with 0.20 Nm^{-1} of spring constant and 13 kHz of resonance frequency) with intermittent indentation into the analysed layers to probe their stiffness. To create such intermittent vertical contact, the sample is subjected to the high ultrasonic frequency vibration ($\approx$ 4.30 MHz) of a 2.70 kHz modulated AC bias applied at a ceramic piezotransducer stage from an Agilent 33500B GEN. This vibration (much higher than the cantilever's resonance frequency) is detected by the tip non-linearly, functioning as an inertially stiff probe with $\approx$ 104 Nm^{-1} of spring constant, resulting in the mechanical response of the contact. The 2.70 kHz modulation is chosen to be lower than the resonance of the cantilever and higher than the 1 kHz feedback cut-off, this way, the amplitude (A_u) is cyclically modulated by the LIA (Stanford Research Systems SR830 DSP) and the tip spans between both the linear and non-linear regimes in a single cycle (depending on the threshold amplitude at which the contact is interrupted). Figure 7.5b represents the normal force (F_N) versus tip/sample distance (Z) UFM curve, where the mentioned A_t limits the amplitude of the sinusoidal modulation. At minimal oscillations (A_0), the low averaged contact force ($F_{N,1}$) causes a negligible cantilever deflection, while, when the amplitude increases up to the threshold value, a jump on the normal force is produced (Fig. 7.5c) [27].

Quantification of the Domains' Young's Modulus

The UFM is conducted on a flat area of the TDBG (Fig. 7.6a) achieving lower contrast of the Moiré superlattice features compared to the electromechanical measurements, suggesting a diminished nanomechanical influence (in agreement with the E-HFM discussion). However, it is still interesting to quantify the elastic properties. Knowing the calibration procedure explained in A.2.4, the UFM map of the area of interest in TDBG is obtained and transformed to Young's modulus, as shown in Fig. 7.6b. The two boundary types are also indicated in the image, however, the SP domain walls separating ABAB and ABCA regions show a much lower contrast compared to the DSP double domain walls, originating from a decreased UFM signal after the domain walls' bifurcations. In the electromechanical measurements, this same wall splitting can be observed, however, the contrast is homogeneous before and after the wall division, indicating the same kind of flexoelectric response for both types of boundaries. Nevertheless, a quantified analysis of the nanomechanical map must be undertaken to understand the different UFM responses in the SP and DSP walls.

Extracting the profile from the cross-plane Young's modulus map (Fig. 7.6c) allows to measure a full width at half maximum value of 12 nm, corresponding to the width of the DSP domain walls. This same profile also quantifies E along the ABAB domains, which results in 40.50 $\pm$ 1.50 GPa, close to the previously stated literature value of 39.51 GPa [28]. The extracted value for the ABCA domains also corresponds to approximately the same Young's modulus, indicative of the similar energy states of both domain types. However, E decreases drastically at the DSP boundaries to 33.29 $\pm$ 0.73 GPa and to 36.92 $\pm$ 0.54 GPa at the SP domain walls (Fig. 7.6d). Therefore, a softening occurs at such domain interfaces, and can be attributed to the incommensurate lower interlayer coupling energy of the atoms at the domain walls together with their increased interatomic distance, agreeing with literature studies

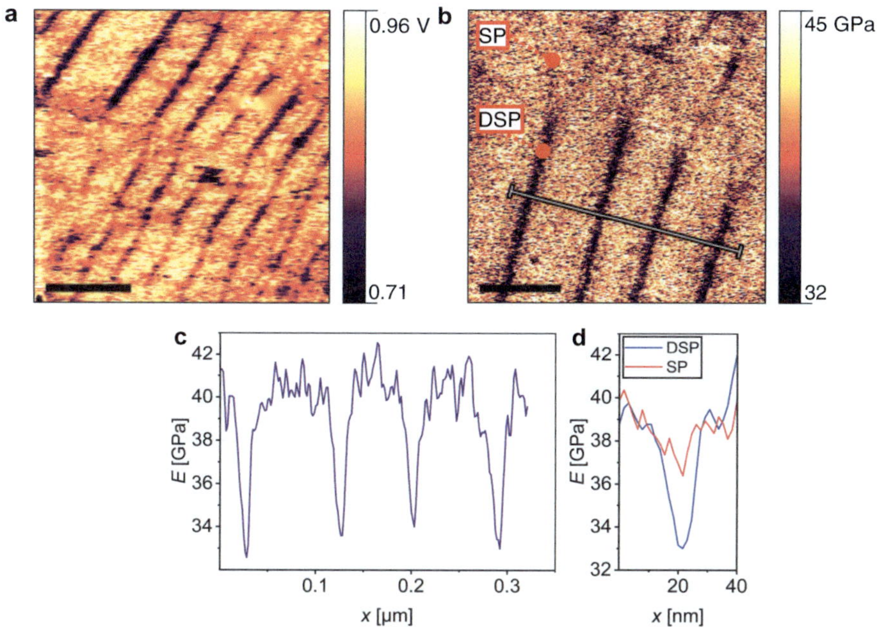

Fig. 7.6 Nanomechanical UFM analysis of TDBG. **a** Original UFM data map (scale bar is 200 nm), **b** magnified UFM map converted to cross-plane Young's modulus of a region in the TDBG superlattice containing single (SP) and double (DSP) domain walls (scale bar is 100 nm), **c** corresponding Young's modulus extracted profile from **b** and **d** comparison between the Young's modulus profiles of a SDW and a DDW

performed on domain walls of ferroelastic and ferroelectric materials. Such investigations also show decreased SP energy at the domain walls related to a softer elastic behaviour [29–31]. If the softening on the DSP is compared to the SP domain walls, one can see that the Young's modulus for the single wall boundaries decreases approximately half of the double wall decrease. This suggests a possible influence of the atomistic reconfiguration on the domain wall nature, however, quantum mechanical simulations at high resolution would have to be performed to confirm such theory.

7.2.4 Force Field Relaxation Simulations of Single Domain Walls

Despite not reaching the required resolution to explain the mentioned atomic reconstruction of the domain walls, force field relaxation simulations can be applied to understand the nanoscale local variations in Young's modulus shown in the previous section [32]. Specifically, a classical force-field model is used determining the intra- and interlayer forces by the optimised Tersoff-Brebber and Kolmogorov-Crespi potentials, respectively, while maintaining the bottom graphene layer flat to mimic the experimental setup and optimising the atomic structure until the force components are smaller than 0.50 meV atom^{-1}.

Hence, the curvature of the interatomic potential energy versus applied strain plot at the relaxed state is computed to obtain the cross-plane Young's modulus of four-layer Bernal-stacked graphene, resulting in a value of 35.50 GPa, similar to the UFM experimental 40.50 GPa value. Then, a local cross-plane Gaussian strain is applied to model the nanomechanical properties of the Moiré superlattice with a twist angle of 0.76°, obtaining the map shown in Fig. 7.7a. In such simulation type, the computation cost is limited by the interlayer angle, rendering impossible to explore smaller twist geometries (where DSPs appear) and, hence, allowing only to compile single domain walls [16].

The obtained map supports the experimental results, showing a higher stiffness for the ABAB and ABCA domains compared to SP walls and ABBC regions. The softening of the single domain walls can also be observed in Fig. 7.7c, with a similar response to the UFM profile. The mechanisms under this SP softening are represented in Fig. 7.7b, where an in-plane misalignment between the two 2LG structures facilitates a vertical displacement compared to the commensurate ABAB regime. Said mechanism is confirmed in Fig. 7.7d, showing the change of interlayer coupling

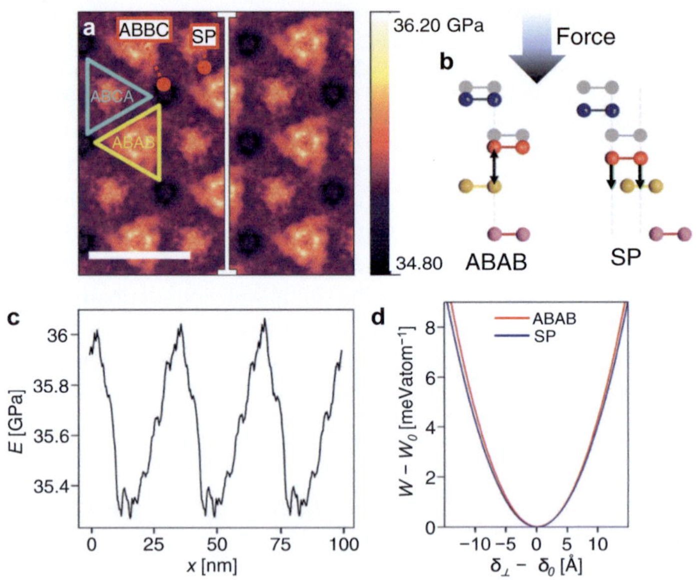

Fig. 7.7 Force field relaxation simulation analysis. **a** Young's modulus force field relaxation modelling map of a TDBG atomic region showing the different stacking orders (scale bar is 40 nm), **b** scheme of the atomic displacement in TDBG ABAB domains and domain walls after an application of an external force, **c** Young's modulus profile extracted from **a** and **d** interlayer coupling energy versus interlayer distance plots in ABAB domains and domain walls. Adapted from Canetta et al. [37]

energy ($W - W_0$) versus distance ($\delta_\perp - \delta_0$) dependence for SP and ABAB regimes, implying a modified response to cross-plane strain and different E.

7.3 Thermomechanical Properties of Buckypapers

7.3.1 Synthesis of Nanotube Networks Forming Buckypapers

A simplified scheme of the fabrication procedure for three different types of BPs based on 1D nanotubes is shown in Fig. 7.8. The synthesis for the MWCNT@BNNT coaxial powder starts with the precursor mixture of vacuum-pretreated ammonia borane powder with inert h-BN filler (0.10 and 1 g, respectively), separating the ammonia borane granules and preventing the global polymerisation of volatile gases emitted upon heating. 3 mg of Nanocyl NC7000™ MWCNT (also pretreated to remove contamination) are spread on a clean quartz slide forming a 2×2 cm^2 flat square, this quartz slide is then inserted into the tube of the atmospheric pressure CVD furnace with a connected flask containing the precursor mixture and purged with argon and hydrogen (2.50%) for 25 min after sealing the chamber. The deposition begins with the alignment of the preheated (90 °C) precursor flask and the quartz slide with the sample heated at 1000 °C. Afterwards, a net carrier 2.50% hydrogen and argon gas flows through the tube and the precursor flask at 400 and 100 SCCM, respectively, leaving the setup for 2 h. Before unsealing, the furnace is shifted downstream, and the sample is cooled to RT for 15 min with the same gas flow.

To synthesise the BPs, 4 g of the purchased MWCNT powder are dispersed in 20 mL of EtOH and sonicated at RT and 20 kHz for 4 h. Next, the dispersion is vacuum filtered on a Cytiva Whatman™ Cyclopore™ polycarbonate membrane

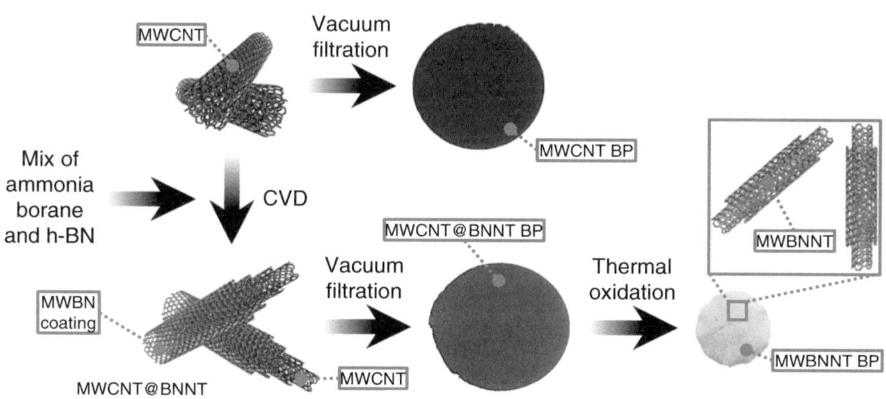

Fig. 7.8 Steps for the fabrication of the three different species of nanotubes and their corresponding buckypapers

with 0.40 μm pore dimension to be subsequently vacuum treated overnight before the BP peeling. The same process can be applied to fabricate MWCNT@BNNT BPs, employing the powder synthesised previously with the coaxial nanotubes. The diameter of the filtration cylinder limits the size of both BPs to 17 mm. Finally, the MWBNNT BP is produced following a sacrificial templating method, starting from a MWCNT@BNNT BP being deposited on a quartz slide. This slide is introduced into the CVD tube to be oxidised under an airflow of 1.80 L min^{-1} at 700 °C for 1 h so that the graphitic components are selectively removed.

7.3.2 Morphology Characterisation

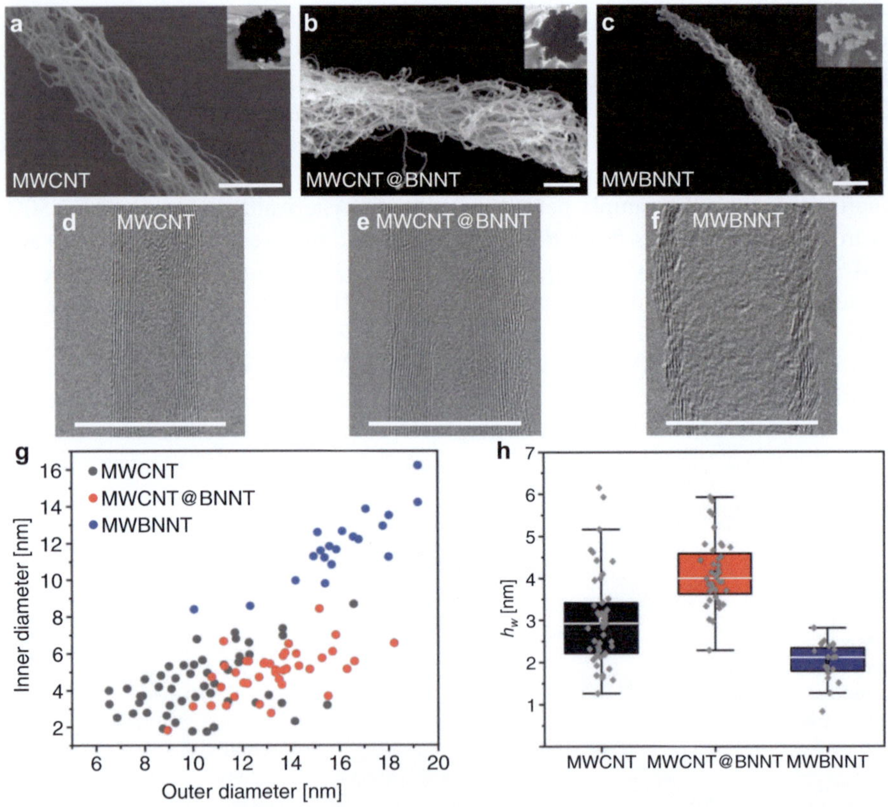

Fig. 7.9 Morphological characterisation of the nanotubes. **a–c** SEM images of the MWCNT, MWCNT@BNNT and MWBNNT nanofibers, respectively (scale bars are 500 nm) (insets show the nanotube respective powders), **d–f** high resolution TEM images of the MWCNT, MWCNT@BNNT and MWBNNT specimens, respectively (scalebars are 20 nm), **g** diameters and **h** wall thicknesses (h_w) of the three different nanotube types

Prior to the thermomechanical characterisation, the synthesised BPs are measured to obtain their precise composition and morphological properties. Their thicknesses ranged from 115 to 140 μm for the MWCNT@BNNT and from 80 to 90 μm for the MWBNNT species. Given the vacuum filtration steps, the resulting BPs present a layered structure with enhanced horizontal plane alignment and flexible behaviour for both MWCNT and MWCNT@BNNT BPs, while MWBNNT presented a brittle stability (due to the reduced length of the nanotubes and entanglement after the thermal oxidation step). Powder samples are extracted from the BPs to be analysed with SEM (Fig. 7.9a–c), where the nanotubes' entanglement can be clearly observed. High resolution TEM identifies the dimensions of individual nanotubes (Fig. 7.9d–f). A hint into the h-BN coating is extracted by the increased outer diameter values of MWCNT@BNNT with respect to the MWCNT, detailed in Fig. 7.9g. MWCNT and the coaxial BP coincide in the range of inner diameter since both share CNT cores, however, such parameter increases in hollow MWBNNT due to the graphitic oxidation. The obtained average wall thicknesses are 3 ± 1 nm, 4.10 ± 0.80 nm and 2 ± 0.60 nm for the MWCNT, MWCNT@BNNT and MWBNNT BPs, respectively (Fig. 7.9h). Additionally, the spectroscopical composition characterisation of the three BP powders can be found in Sect. A.2.5.

7.3.3 Piercing Scanning Thermal Microscopy

The thermal conductivity of such BPs is a subject of interest due to their prospective applications, but their elevated porosity and softness create a major challenge to perform such heat transport measurement with methods that require the deposition of additional structures on the BPs, e.g., thermoreflectance or 3ω-method. One could apply standard SThM as a better alternative to the mentioned methods to measure said heat parameter, however, the lack of well-defined tip/sample contact of the BPs requires a different approach. Such system is called piercing scanning thermal microscopy (pSThM) and is developed here for the first time. It encompasses the probing of the heat conductance into the BPs depending on the penetration depth of the tip (ΔZ), thus relating the heat transport with the mechanical stability of the thin films (Fig. 7.10a). Such thermal conductance (G_{BP}) is given by the next expression derived from the Fourier's law:

$$G_{BP} = \frac{\Theta}{\Delta T} = \alpha_{BP} 2\pi a_{tip} \Delta Z \kappa_{BP} = G_{nc} \frac{\partial T}{\Delta T} \quad (7.3)$$

where Θ is the heat current into the sample and ΔT is the temperature difference between the tip and the BPs. As seen in the same equation, a direct proportionality also allows to convert the thermal conductance into thermal conductivity (κ_{BP}) taking into consideration the cylindrically approximated volumetric perimeter of the tip/sample contact ($2\pi a_{tip} \Delta Z$) and a coefficient (α_{BP}) with m^{-1} dimension in the order of the inverse diameter of the probe and identical for all the BPs [33]. The employed

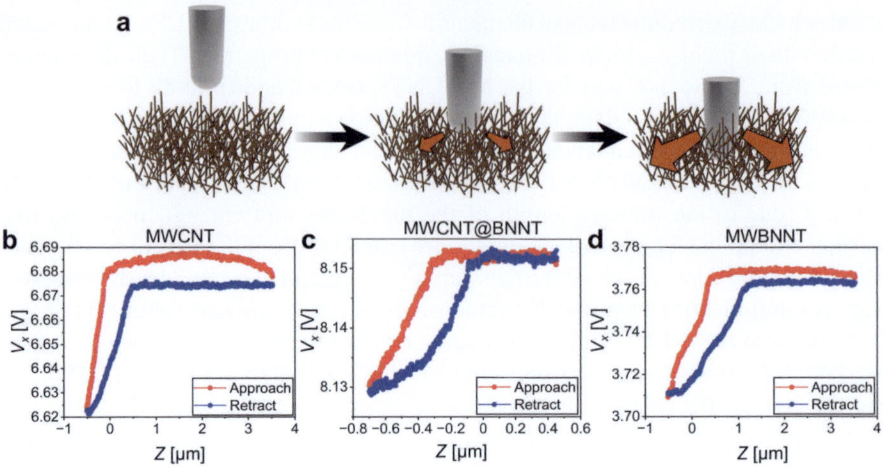

Fig. 7.10 pSThM functioning and response. **a** Scheme of the pSThM tip indenting into the nanotube network with increasing heat transfer represented by the orange arrows, **b** MWCNT, **c** MWCNT@BNNT and **d** MWBNNT approach and retract curves showing dependence between the thermal signal and the penetration depth

HV SThM equipment is similar to the one described in Sect. 4.2, but, instead of raster scanning, the tip (Bruker VITA-HE-NANOTA-200) performs an approach and retract movement indenting (or "piercing") into the sample while it is powered from the BRG at 400 mV (91 kHz) of AC and 600 mV of DC bias, resulting in an approximate 40 K of initial tip/sample temperature excess. When the tip is out of contact, the calculated heat conductance of the probe to its base is $G_{nc} = 1.87 \times 10^{-5}$ W K^{-1}, while, in the contact regime, the G_{BP} is added to the heat conductance, resulting in a probe temperature drop of $\Delta T - \partial T$, given its constant supplied power and $\partial T \ll \Delta T$ [34].

Figure 7.10b–d show the resulting thermal signal approach and retract curves of the pSThM for each BP type. An absence of the typical tip/surface jump is observed, indicating that the sample cannot be standardised as a stiff solid [35]. Moreover, a linear decrease of the thermal signal once in the contact regime reflects the linear increase of the heat flow rate into the sample depending on the penetration depth. Both approach and retract plots have similar dependencies, only differing in that the retracts expand over larger distances since the BP fibres release the tip gradually. The deflection of the cantilever is negligible compared to the movement of the tip, given the extremely soft behaviour of the BPs and the stiff probe, understanding the tip displacement as the penetration depth (ΔZ).

The derivation of Eq. (7.3) per penetration depth results in a direct proportionality of the experimentally measured $\partial T \Delta Z^{-1} \Delta T^{-1}$ to the κ_{BP}. Figure 7.11 shows the experimental values for the three types of BPs, where the lowest thermal conductivity of MWBNNT BP is observed (30% less than the MWCNT and MWCNT@BNNT BPs). Such low heat transfer behaviour is due to the partial oxidation and reduced

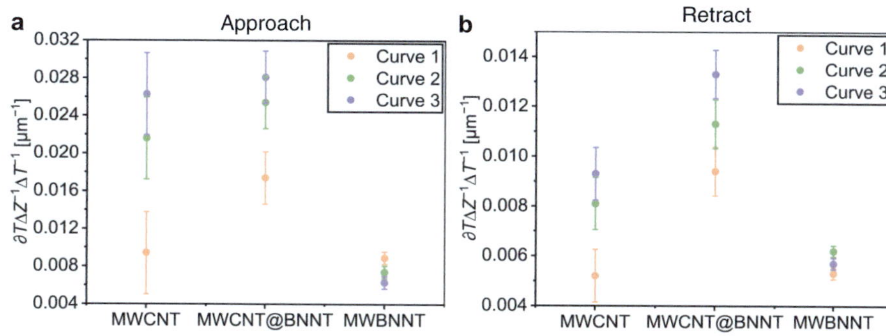

Fig. 7.11 Plots showing the derivative of the relative change of the probe's temperature per increment of penetration depth for **a** approach and **b** retract pSThM curves in the studied nanotubes

crystallinity and length of the nanotubes after the thermal oxidation fabrication step (as seen in Sect. A.2.5). Importantly, the conductivity of the MWCNT@BNNT BP is the highest among the three samples (14% higher than the pristine MWCNT BP). Specific values of κ_{BP} can be provided assuming a good thermal contact between the tip and the BPs, obtaining 1.64 ± 0.50, 1.87 ± 0.50 and 0.76 ± 0.20 W m^{-1} K^{-1} for the MWCNT, MWCNT@BNNT and MWBNNT BPs, respectively. It must be noted that these values correspond to the apparent or effective thermal conductivities of the layers, given their porous structure contribution to heat transport, converting the materials to heterogeneous specimens. Nevertheless, the obtained κ_{BP} values are lower than those of individual nanotubes, suggesting an interfacial thermal contact resistance at the nanotubes' junctions and a misalignment of the 1D structures [36]. The achieved κ_{BP} increase for the MWCNT@BNNT BP proves the effectivity of the coaxial h-BN coating on the MWCNT for introducing additional heat transfer pathways along the 1D direction while electrically shielding the graphitic core.

7.4 Summary

The interplay between thermal, mechanical and electrical properties has been characterised in specific nanoscale systems for the first time [37, 38]. The first part encompasses the usage of three different SPM techniques in order to image the Moiré pattern of TDBG and obtain its nanomechanical and electromechanical response. PFM allows to observe the presence of non-uniform strain between the TDBG layers, appearing as elongated triangular domains in the Moiré superlattice. Such strain causes a flexoelectric phenomenon along the domain single and double boundaries, allowing for local modification of the electromechanical properties. The dynamic relaxation of the TDBG is also quantified by means of E-HFM mapping, resulting in an ~ 8 ns phase delay originating mostly from the flexoelectric response. Moreover, the Young's modulus of such structure is locally quantified using UFM, showing a decrease of 7.21 GPa at the DSP walls. The variation of Young's modulus is further interpreted with force field relaxation simulations agreeing with the experimental data. Overall, the results exposed here confirm the suitability of nano- and electromechanical SPM techniques to locally extract quantified data of the dynamic and stability behaviour of twisted vdW materials, opening the door to device durability enhancements and reduced local geometry for high efficiency electromechanical sensing or activation.

The second part of the chapter is based on the fabrication and thermomechanical analysis of nanotube networks forming BPs. Thanks to a scalable and accessible atmospheric pressure CVD, three types of BPs are successfully synthesised, namely MWCNT, MWCNT@BNNT and MWBNNT samples. The newly developed pSThM methodology allows to measure the soft nature of the BPs and their heat conductivities, suggesting a positive effect of coating the CNT core with BNNT to increase such conductivity by 14% and electrically shielding the transverse direction. Such coaxial structure could therefore be applied to systems requiring directional electrical transport and enhanced heat conductivity. Also, MWBNNT BP shows a reduced thermal conductivity value given its reduced crystallinity acting as an attraction source for oxidation states, which could potentially be employed in carbon dioxide or water adsorption applications.

References

1. Cao, Y., et al. (2018). Unconventional superconductivity in magic-angle graphene superlattices. *Nature, 556*, 43–50.
2. Novelli, P., Torre, I., Koppens, F. H. L., Taddei, F., & Polini, M. (2020). Optical and plasmonic properties of twisted bilayer graphene: Impact of interlayer tunneling asymmetry and ground-state charge inhomogeneity. *Physical Review B, 102*, 125403.
3. Sharpe, A. L., et al. (2019). Emergent ferromagnetism near three-quarters filling in twisted bilayer graphene. *Science, 365*, 605–608.
4. Liu, A., & Peng, Q. (2018). A molecular dynamics study of the mechanical properties of twisted bilayer graphene. *Micromachines, 9*, 1–11.
5. Zheng, S., Cao, Q., Liu, S., & Peng, Q. (2019). Atomic structure and mechanical properties of twisted bilayer graphene. *Journal of Composites Science, 3*, 1–8.
6. Zhang, J., & Zhao, J. (2013). Mechanical properties of bilayer graphene with twist and grain boundaries. *Journal of Applied Physics, 113*, 043514.
7. Alden, J. S., et al. (2013). Strain solitons and topological defects in bilayer graphene. *Proceedings of the National Academy of Sciences of USA, 110*, 11256–11260.
8. Woods, C. R., et al. (2014). Commensurate–incommensurate transition in graphene on hexagonal boron nitride. *Nature Physics, 10*, 451–456.
9. Liu, J., et al. (1998). Fullerene pipes. *Science, 280*, 1253–1256.
10. Endo, M., et al. (2005). 'Buckypaper' from coaxial nanotubes. *Nature, 433*, 476–476.
11. Xia, Q., Zhang, Z., Liu, Y., & Leng, J. (2020). Buckypaper and its composites for aeronautic applications. *Composites, Part B: Engineering, 199*, 108231.
12. Chang, C. W., et al. (2006). Isotope effect on the thermal conductivity of boron nitride nanotubes. *Physical Review Letters, 97*, 085901.
13. Schulz, M. J., Shanov, V. N., & Yin, Z. (2013). *Nanotube superfiber materials* (1st ed.). William Andrew.

14. Chen, X., Dmuchowski, C. M., Park, C., Fay, C. C., & Ke, C. (2017). Quantitative characterization of structural and mechanical properties of boron nitride nanotubes in high temperature environments. *Science and Reports, 7*, 11388.
15. Saito, R., Fujita, M., Dresselhaus, G., & Dresselhaus, M. S. (1992). Electronic structure of chiral graphene tubules. *Applied Physics Letters, 60*, 2204–2206.
16. Halbertal, D., et al. (2021). Moiré metrology of energy landscapes in van der Waals heterostructures. *Nature Communications, 12*, 242.
17. Li, Y., et al. (2021). Unraveling strain gradient induced electromechanical coupling in twisted double bilayer graphene Moiré superlattices. *Advanced Materials, 33*, 2105879.
18. Yoo, H., et al. (2019). Atomic and electronic reconstruction at the van der Waals interface in twisted bilayer graphene. *Nature Materials, 18*, 448–453.
19. McGilly, L. J., et al. (2020). Visualization of Moiré superlattices. *Nature Nanotechnology, 15*, 580–584.
20. Hattendorf, S., Georgi, A., Liebmann, M., & Morgenstern, M. (2013). Networks of ABA and ABC stacked graphene on mica observed by scanning tunneling microscopy. *Surface Science, 610*, 53–58.
21. Kim, K., et al. (2017). Tunable Moiré bands and strong correlations in small-twist-angle bilayer graphene. *Proceedings of the National Academy of Sciences of USA, 114*, 3364–3369.
22. Shu, L., et al. (2019). Flexoelectric materials and their related applications: A focused review. *Journal of Advanced Ceramics, 8*, 153–173.
23. Cuberes, M. T. (2009). Intermittent-contact heterodyne force microscopy. *Journal of Nanomaterials, 2009*, 762016.
24. Cuberes, M. T., Assender, H. E., Briggs, G. A. D., & Kolosov, O. V. (2000). Heterodyne force microscopy of PMMA/rubber nanocomposites: Nanomapping of viscoelastic response at ultrasonic frequencies. *Journal of Physics D: Applied Physics, 33*, 2347–2355.
25. Wang, J., Wang, L., Xing, Y. H., Liu, T. X., & Zhang, S. T. (2018). Mechanism of large tunable thermal transport in graphene with oxygen functional groups. *Journal of Applied Physics, 124*, 175108.
26. Cong, X., et al. (2019). Probing the acoustic phonon dispersion and sound velocity of graphene by Raman spectroscopy. *Carbon, 149*, 19–24.
27. Dinelli, F., Albonetti, C., & Kolosov, O. V. (2011). Ultrasonic force microscopy: Detection and imaging of ultra-thin molecular domains. *Ultramicroscopy, 111*, 267–272.
28. Cho, J., Luo, J. J., & Daniel, I. M. (2007). Mechanical characterization of graphite/epoxy nanocomposites by multi-scale analysis. *Composites Science and Technology, 67*, 2399–2407.
29. Jiang, L., et al. (2018). Manipulation of domain-wall solitons in bi- and trilayer graphene. *Nature Nanotechnology, 13*, 204–208.
30. Lee, W. T., Salje, E. K. H., & Bismayer, U. (2003). Domain wall diffusion and domain wall softening. *Journal of Physics: Condensed Matter, 15*, 1353.
31. Stefani, C., et al. (2020). Mechanical softness of ferroelectric 180° domain walls. *Physical Review X, 10*, 041001.
32. Leven, I., Maaravi, T., Azuri, I., Kronik, L., & Hod, O. (2016). Interlayer potential for graphene/h-BN heterostructures. *Journal of Chemical Theory and Computation, 12*, 2896–2905.
33. Rohsenow, W. M., Hartnett, J. R., & Cho, Y. I. (1998). *Handbook of heat transfer* (3rd ed.). McGraw-Hill Education.
34. Tovee, P., Pumarol, M., Zeze, D., Kjoller, K., & Kolosov, O. (2012). Nanoscale spatial resolution probes for scanning thermal microscopy of solid state materials. *Journal of Applied Physics, 112*, 114317.
35. Spiece, J., et al. (2018). Improving accuracy of nanothermal measurements via spatially distributed scanning thermal microscope probes. *Journal of Applied Physics, 124*, 015101.
36. Han, Z., & Fina, A. (2011). Thermal conductivity of carbon nanotubes and their polymer nanocomposites: A review. *Progress in Polymer Science, 36*, 914–944.
37. Canetta, A., et al. (2023). Quantifying the local mechanical properties of twisted double bilayer graphene. *Nanoscale, 15*, 8134–8140.

38. Jones, R. S., et al. (2023). Thermal conductivity of carbon/boron nitride heteronanotube and boron nitride nanotube buckypapers: Implications for thermal management composites. *ACS Applied Nano Materials, 6*, 15374–15384.

Chapter 8
Conclusions and Perspectives

Abstract The last chapter of the book presents the main conclusions drawn from the experimental findings and methodological developments explored throughout this work. It highlights the successful fabrication and characterisation of 2D materials and nanostructures, the demonstration of anisotropic heat transport and the optimisation of thermoelectric performance through interfacial engineering. Local modulation of the Seebeck coefficient and pioneering electro- and thermomechanical characterisation of twisted and unidimensional systems are also summarised. Limitations related to spatial resolution, diffusive assumptions and temperature range are discussed. Finally, future research directions are proposed, emphasising the need for quantum-level modelling, cryogenic SPM and atomic-scale transport studies.

2D materials and their heterostructures have proven to be resourceful tools for the controlled enhancement of performance in modern devices. Their properties have been extensively measured over recent years, with a plethora of groundbreaking discoveries enlarging the materials science field. However, such investigations tend to focus more towards specific physical areas like electronic, magnetic or optical behaviour, demoting the contributions related to other phenomena such as thermal transport, nanomechanics or thermoelectricity. Given the linkage between all types of properties in solid materials, it is crucial to understand the nature of this interplay. However, the highly anisotropic properties of vdW structures often lack appropriate characterisation methodologies.

For this purpose, this thesis aims to improve the understanding of the thermal and mechanical properties of two-dimensional systems and their intrinsic interrelation between each other as well as with electronic transport, with a crucial focus on thermoelectricity and electromechanics. This is done starting from the origin by improving the synthesis of such types of structures employing state of the art techniques and their characterisation via SPM nanoscale resolutive approaches capable of providing a profound insight into the materials' characteristics. The main outcomes obtained in this work as well as their related future improvement pathways are described in the two sections below.

S. Gonzalez-Munoz, *Interplay of Nanoscale Physical Properties in 2D Materials and Heterostructures*, Springer Theses, https://doi.org/10.1007/978-3-032-14271-9_8

8.1 Key Results

Optimised Procedures for Nanoscale Sample Fabrication

In the search for combining quality with time efficiency and reproducibility, multiple synthesis methodologies for the creation of 2D structures are described in Chap. 3. Starting from the steps undertaken to mechanically exfoliate thin flakes of four types of commonly used 2D materials on standard substrates and highlighting the importance of pretreatment of the supporting layers. Also, three different variations of micromanipulated stamping are tested to form vdW homo- and heterostructures, resulting in the PDMS classical procedure as the most time-effective while keeping low polymeric contamination levels. Patterning and metal deposition are also employed here to create nanoscale devices based on 2D materials, with a novel method of indium needle deposition functioning as electrodes. Lastly, BEXP is described as an improved alternative to FIB capable of providing access to the inner layers of the studied materials with minimal surface damage and expanded measurable areas. All the fabrication techniques described here represent tunable and versatile options for the nanoscale synthesis of samples intended to be further analysed.

Anisotropic Heat Transport in Two-Dimensional Systems

Due to their geometry and binding configuration, 2D materials tend to show very different behaviour when analysed along their basal plane vs. the perpendicular direction. This phenomenon also affects the heat transport of such structures. Thus, Chap. 4 focuses on the quantitative anisotropic thermal conductivity measurements of different vdW structures employing a novel methodology with nanoscale resolution named xSThM. This characterisation technique includes an analytical modelling (validated with FEA simulations) that allows to extract both in-plane and cross-plane thermal conductivity values and evaluate the upper limits of the interfacial thermal insulance between the 2D layers and their substrates, eliminating the requirement of knowing the tip / sample thermal resistance. Moreover, the obtained thermal conductivities for InSe and perovskite semiconductor materials result in anomalously low values, opening the door for the application of such specimens to high efficiency thermoelectrics or optoelectronics.

Enhanced Thermoelectric Performance of VdW-Based Structures

The field of thermoelectricity is a promising candidate for renewable energy conversion, however, it currently lacks the options of core materials that can deliver proper efficiency, given the strong cross-linkage between heat and electronic transport. In Chap. 5, two types of bulk structures based on two-dimensional configurations are studied with the purpose of enhancing the actual limits of thermoelectric efficiency. To do so, the individual evaluation of the structural, electronic and thermal characteristics of Sb_2Te_3 / MoS_2 superlattices of different thicknesses and Sb_2Te_3 / $AgSbTe_2$ hybrid nanocomposites with varied doping concentrations is undertaken. Enhanced power factors are found thanks to the selective screening of the charge carriers at the synthesised interfaces. Such junctions also aid in the reduction of the thermal transport along the anisotropic directions due to phonon scattering. Specifically, the modification of the crystallinity of nanocomposites and the introduction of low thermal conductivity MoS_2 in the superlattices allow to precisely engineer the electronic and phononic transport, achieving a record-high value of $ZT = 2.26$ close to room temperature. These results lead to the beneficial application of such types of structures in thermoelectric management platforms and to the possible enhancement of their efficiency by the combination of other suitable materials.

Local Thermoelectricity Modulation via Geometrical Patterning

As described in Chap. 5, the thermoelectric characterisation is generally performed along the entire structure. However, the nanoscale nature of 2D materials allows to precisely modulate many of their properties at a local spatial distribution, including the thermopower or Seebeck coefficient. Chapter 6 serves this purpose by patterning nanoscale constrictions on two-dimensional devices to ultimately characterise the local thermoelectric variations arising in such geometrical features. The asymmetric and hole-array constrictions in encapsulated graphene and $SnSe_2$ are characterised for the first time via STGM, probing the dependence of the thermovoltage with respect to the composition, gate doping, temperature, constriction geometry and constriction influence of the samples. The nanoconstrictions are proven to create a spatial-dependent thermovoltage with inversed polarity due to the close scattering boundaries, which modify the electronic mean free path. Moreover, backgate doping and the variation of the temperature gradient are found to be capable of modulating the nature of such thermovoltage in graphene, providing additional degrees of freedom for the thermoelectric local management in vdW materials. Interestingly, nearby constrictions function as thermovoltage destructive interference sources in the $SnSe_2$ rectangular hole arrays, whereas expanded inverse polarity thermovoltage gradients are observed at the sides of such patterns, denoting the importance of the Seebeck decay length in such layered systems.

Electro- and Thermomechanical Characterisation of Nanoscale Structures

The areas of twistronics and unidimensional structures have gained increasing importance in the scientific community in recent years, however, due to their novelty, the amount of specific characterisation is still in the early phases, especially when talking about the mechanical properties and their synergy with thermal and electronic transport. Aiming towards an improved understanding of such types of behaviours, a quantification of the nano-, electro- and thermomechanical properties of low-dimensional structures with nanoscale resolution is conducted in Chapt. 7 for the first time. The chapter is divided into two parts. Firstly, three different frequency-modulated SPM techniques are employed to probe the nanomechanical and piezoelectric response of TDBG. This allows to extract images of the formed Moiré pattern due to the intrinsic interlayer strain, to observe a flexoelectric behaviour at the domain boundaries and to calculate the dynamic electromechanical relaxation of the stacked structure (in the order of few ns). Also, the Young's modulus is locally quantified, showing a decrease of 7.21 GPa at the domain walls. Finally, the fabrication and thermomechanical analysis of BPs based on carbon and boron nitride nanotubes is assessed in the second part of the chapter. Here, an enhanced atmospheric pressure CVD is used to synthesise three different types of BPs, and a novel pSThM method is used to obtain their thermal conductivities, linked with their soft mechanical stability. Such measurements indicate a heat transport improvement with the h-BN coating of CNT, while isolated BNNTs show a tendency of oxygen absorption that could be employed in filtering applications.

8.2 Perspectives

While delving into multiple scientific advancements related to nanoscale materials in this thesis, the inherent vastness of this field guarantees an ever-expanding landscape, continuously unveiling novel horizons. Therefore, starting from the discussed methodologies and findings, deeper insights into the mechanical, electrical and thermal properties of 2D materials as well as their interplay could be achieved by further investigation employing advanced scanning probe techniques.

Heat transport anisotropy has proven to be influenced by multiple parameters, including the composition, the substrate proximity and the thickness of the layers. However, the measurements performed here are limited to the diffusive transport regime, disregarding the thermal transport in minimal thickness materials with higher mean free paths. For their characterisation, a ballistic transport model would be needed, opening the door for the combination of such simulations with a higher resolutive SPM method capable of measuring independent monolayers. This could be performed by improving the tip's apex radius with the addition of CNT, for instance. Also, a rougher surface is observed in the xSThM of mechanically exfoliated flakes, having an effect on the extracted thermal properties that, although being minimal, must be further studied and compensated. This could be performed by a post-annealing process under vacuum conditions to flatten such bulged surfaces, and it could be compared to the same analysis on further semiconductor materials supported by different substrates (or even suspended) via SPM to deconvolute its origin and any possible ballistic contribution.

Furtherly, the nanocomposite samples based on 2D configuration show interesting thermoelectric efficiency values, however, they are only achievable at high temperatures. With the current trend of enhanced *ZT* structures at RT, it is needed to decrease the heat range at which such nanocomposites show a competitive energy conversion rate. For this, future pathways must be both the intrinsic characterisation of the in-plane and cross-plane thermal and electronic properties and the formation of novel nanocomposites that could own carrier filtering mechanisms (as the ones exposed here) at lower temperatures. Given that the xSThM methodology extracts such anisotropic low thermal conductivity values, this could be a starting point for novel anisotropic electronic measurements based on the SPM of TMD nanocomposites.

Following with thermoelectricity, an extensive toolkit for modification of the local distribution of the Seebeck coefficient is presented here. Nevertheless, the nature of the scattering influence on both thermal and electronic transport must be understood at a quantum level. In such regard, molecular and even atomic precise simulations would be needed, perhaps accompanied by ultrasensitive SPM techniques capable of measuring such type of carrier distribution influence. Scanning tunnelling and photothermal-induced resonance microscopies could be employed at ultrahigh vacuum and cryogenic temperatures to accomplish such tasks. However, the latter is still in the early stages of development.

Lastly, electromechanical behaviour in twisted layers shows a hidden potential with several underlying promising outcomes. However, due to the atomic scales involved, the same resolution problem is faced, requiring accompanying the SPM techniques employed for the characterisation of quantum strained systems with precise simulations or the comparison with results obtained from other advanced frequency-modulated microscopy techniques. Moreover, this could open the door for further interesting applications, like the local ferroelectric switching of piezoelectric materials such as α-In_2Se_3 with PFM for quantum activation and sensing systems.

All of these developments require continuous investigation, which is being continued in the Lancaster University Physics Department with the scope of reaching such nanoscale horizons. In the end, it has been proven that are, in fact, the tiny elements the ones capable of walking gigantic advancement steps.

Appendix

A.1 Contributed Oral Presentations and Posters

- Castanon, E. G.[1], Niblett, A., Gonzalez, S., Hamer, M. J., Zultak, J., Kazakova, O. & Kolosov, O. V. STGM studies of charge symmetry in the geometrically modulated Seebeck coefficient in encapsulated graphene nanoconstrictions. *Microscience Microscopy Congress (MMC) 2021* (Virtual). **Poster and flash talk presentation**.
- Agarwal, K.[1], Gonzalez-Munoz, S., Ahmad, M., Mehta, B. R. & Kolosov, O. V. Thermal transport in enhanced thermoelectric performance high FOM Sb_2Te_3/ MoS_2 heterostructure. *Graphene Week 2021* (Virtual). **Poster and flash talk presentation**.
- Castanon, E. G.[1], Hamer, M. J., Gonzalez, S., Niblett, A., Zultak, J., Gorbachev, R., Kazakova, O. & Kolosov, O. V. Thermoelectric voltage modulation via backgate doping in graphene nanoconstrictions studied with STGM. *Graphene Week 2021* (Virtual). **Poster and flash talk presentation**.
- Agarwal, K.[1], Gonzalez-Munoz, S., Ahmad, M., Spièce, J., Mehta, B. R. & Kolosov, O. V. Quantitative nanoscale electrical and thermal transport studies in enhanced thermoelectric performance Sb_2Te_3/MoS_2 multilayer sample. *2022 MRS Spring Meeting & Exhibit* (Virtual). **Oral presentation**.
- Kolosov, O. V.[1], Castanon, E., Spièce, J., Evangeli, C., Hamer, M., Gonzalez-Munoz, S., Niblett, A., Zultak, J., Kazakova, O. & Gorbachev, R. Electrical vs spatial symmetry in geometrically defined single-material graphene thermoelectric devices. *2022 MRS Spring Meeting & Exhibit* (Virtual). **Oral presentation**.
- Canetta, A.[1], Spièce, J., Gonzalez-Munoz, S., Nguyen, V. H., de Crombrugghe, P., Agarwal, K., Hong, Y., Mohapatra, S., Watanabe, K., Taniguchi, T., Ribeiro-Palau, R., Charlier, J. C., Kolosov, O. & Gehring, P. Local nanomechanical properties

[1] Main presenter.

S. Gonzalez-Munoz, *Interplay of Nanoscale Physical Properties in 2D Materials and Heterostructures*, Springer Theses, https://doi.org/10.1007/978-3-032-14271-9

in twisted double bilayer graphene. *Graphene 2022* (Aachen, Germany). **Oral presentation**.

- Kolosov, O.[1], Castanon, E. G., Hamer, M. J., Gonzalez, S., Zultak, J., Kazakova, O. & Gorbachev, R. Carriers polarity vs spatial symmetry in "geometrical thermoelectricity" phenomenon in the encapsulated graphene thermoelectric devices. *Graphene Week 2022* (Munich, Germany). **Poster presentation**.
- Ahmad, M.[1], Agarwal, K., Gonzalez-Munoz, S., Ghosh, A., Kolosov, O. V. & Mehta, B. R. Interface effects on electron and phonon transport and large enhancement in thermoelectric properties in Sb_2Te_3/MoS_2 multilayer. *E-MRS 2022 Fall Meeting* (Warsaw, Poland). **Contributed talk**.
- Agarwal, K., Gonzalez-Munoz, S., Ghosh, A., Ahmad, M., Mehta, B. R. & Kolosov, O. V.[1] Thermal transport and outstanding thermoelectric properties of 2D-3D Sb_2Te_3 based heterostructures with ZT > 2.0. *2022 MRS Fall Meeting & Exhibit* (Boston, United States). **Oral presentation**.
- Kolosov, O. V.[1], Gonzalez-Munoz, S. & Agarwal, K. Direct measurements of anisotropic thermal transport in 2D materials and heterostructures. *Graphene 2023* (Manchester, United Kingdom). **Poster presentation**.
- Chen, Y., Gonzalez-Munoz, S., Wu, W. & Kolosov, O.[1] From molecular-scale electrical double-layer structure to 3D nano-rheology properties of solid electrolyte interphase. *Microscience Microscopy Congress (MMC) 2023* (Manchester, United Kingdom). **Contributed talk**.
- Medapalli, R.[1], Agarwal, K., Gonzalez-Munoz, S., Kafanov, S., Jarvis, S., Mikhaylovskiy, R. & Kolosov, O. Sub-wavelength focusing of mid-IR light using metal/diamond/metal campanile probe for ultra-broadband SPM. *Microscience Microscopy Congress (MMC) 2023* (Manchester, United Kingdom). **Poster presentation**.
- Kolosov, O.[1], El Sachat, A., Evangeli, C., Xiao, P., Gonzalez, S., Finch, S., Castanon, E., McCann, E., Kazakova, O. & Gorbachev, R. Scanning thermal microscopy (SThM) of vdW materials for discovery of non-conventional thermoelectricity. *AFM & SPM Meeting 2024* (Durham, United Kingdom). **Submitted talk**.
- Medapalli, R.[1], Cottam, N., Agarwal, K., Gonzalez-Munoz, S., Mikhaylovskiy, R., Kafanov, S., Jarvis, S., Patane, A. & Kolosov, O. Sub-wavelength focusing of mid-IR light using metal/diamond/metal campanile probe for ultra-broadband SPM. *AFM & SPM Meeting 2024* (Durham, United Kingdom). **Submitted talk**.

A.2 Additional Methodology

A.2.1 General Concepts of Exfoliation

A common feature of 2D materials is their interlayer weak vdW bonding, which allows them to separate from their mother bulk crystal (Fig. A.1) via exfoliation. However, this interlayer attachment strength varies from structure to structure due to different arrangement types, resulting in materials that are easier to exfoliate than others. Also, the in-plane and edge defects as well as the covalent bonds play a major role in the final lateral size and thickness of the separated flakes.

To get an accurate idea of how challenging it is to separate vdW layers from their 3D counterpart, one can look at the interlayer binding energy, which defines the minimum energy required to split such structures. The theoretical pristine structure has the minimum local binding energy homogeneously distributed within the basal layer, meaning that each individual atom will bind with the same strength to its cross-plane adjacent layer (only depending on the composition and commensurate state dictated arrangement), however, interplane defects like intercalation, vacancies or dislocations common in the growth of crystals cause a change in the binding energy of the same layer, promoting to an easier division of the weaker area from the basal plane as well as a shear force between this area and the tightly bonded region, which usually remains on the bulk structure. This paper-like rupture originates from the lateral edges of such layers, specifically from defects in the form of oxygen or hydrogen groups' substitutions or dislocations, promoting an easier strip of the inner in-plane covalent bonds along the continued defects or the preferred A-C or Z-Z tearing direction [1, 2]. Thus, when a layered material has a higher density of edge defects, meaning a greater number of rupture sources, the lateral areas are more rapidly reduced in

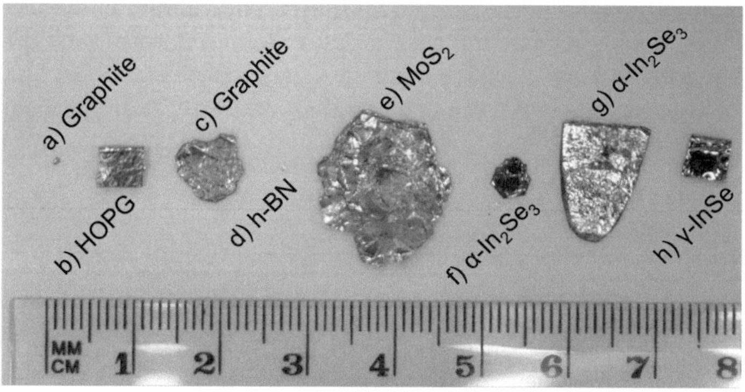

Fig. A.1 Bulk crystals employed for this thesis' exfoliation. **a** HQ graphene natural graphite, **b** HQ graphene highly oriented pyrolytic graphite (HOPG), **c** ProGraphite natural graphite giant flake, **d** HQ graphene hexagonal boron nitride, **e** HQ graphene 2H-MoS_2 natural, **f** HQ graphene In_2Se_3 (alpha phase), **g** α-In_2Se_3 and **h** γ-InSe synthesised at the University of Nottingham and the National Academy of Sciences of Ukraine by Dr. Zakhar R. Kudrynskyi and Prof. Zakhar D. Kovalyuk

size after fewer exfoliations. If defects are disregarded, it is still necessary to know the intralayer strength against rupture that 2D materials intrinsically have. The shear modulus (Γ_S) takes into account such resistance to tearing forces, affecting the lateral sizes of the exfoliated flake as well as the preferential rupture direction. To summarise, Table A.1 depicts the interlayer binding energies and the shear modulus of the four exfoliated 2D materials in this thesis, being graphene the easiest one to separate from the bulk crystal and the most probable to result in bigger flakes due to its low binding energy and high shearing modulus.

Interlayer binding energy, in-plane shearing modulus and edge defect concentration constitute the paramount mechanical factors affecting the efficiency of the exfoliation process, however, they are intrinsic to the material and cannot be conveniently controlled and adapted during the exfoliation process. On the other hand, other non-substrate-related parameters can be macroscopically modulated to yield large areas of thin flakes, engineering the quality of the exfoliation.

For instance, the type of tape employed can be beneficial if it has a low–medium level of tackiness, but it can diminish the quality of the procedure if it attaches firmly to the flakes since many of them will remain on the tape after the substrate deposition, with the added that high attachment tape leaves a great amount of glue residue on the substrate (Fig. A.2) [8, 9]. The amount of pressure applied on the deposition steps and its direction also contributes to the exfoliation yield, being preferential to apply maintained medium pressure perpendicularly to the interacting faces or soft pressure in a singular dragging direction parallel to the substrate's surface [8, 9]. This dragging motion is also beneficial for removing the air bubbles trapped within the interacting faces. A backside heat source is also another powerful tool for evaporating water molecules and increasing the flake/substrate contact area and strength [10, 11]. The tape retraction speed and its angle are critical to achieving the desired deposition rate, with the suitable options being medium–high speed and a 90–120° angle in the flake pick-up oppositely to the slow speed and 10–45° angle for the flake drop-down [8]. Lastly, other factors such as humidity or environmental pollution have a major effect on the yield and the final properties of the exfoliation, with some 2D materials like InSe being much more sensitive to air particle adsorption and fast oxidation. To maintain the pristine properties of such exposed materials, their encapsulation and

Table A.1 Exfoliation-related properties of different 2D materials

Material	Binding energy [eV atom^{-1}]	Γ_S [GPa]	References
Graphene	0.06	280	[3–7]
h-BN	0.07	114.70	
MoS_2	0.23	32	
γ-InSe	0.13	11.80	

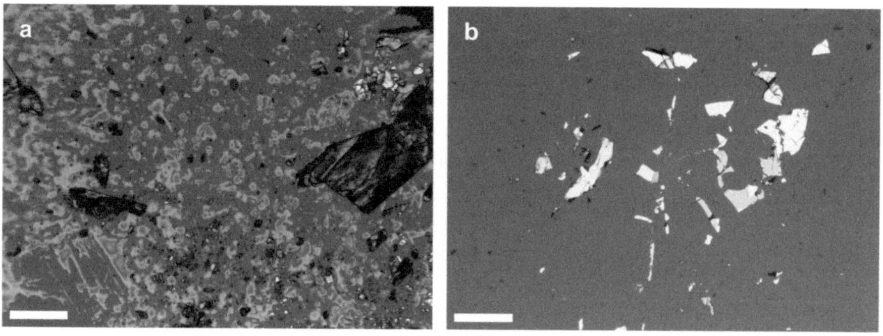

Fig. A.2 Comparison between exfoliation performed with **a** high tackiness Gel-Pak® Gel-Film® PF-6.5-X8 and **b** medium tackiness Gel-Pak® Gel-Film® PF-6.5-X4. Notice the amount of adhesive residue that the high attachment tape leaves on the substrate. Scale bars are 200 μm

isolation between stable layers, such as h-BN [12], can be considered. Another option is to perform such a process under a temperature, pollution and humidity-controlled environment such as a cleanroom or a glovebox filled with an inert gas [8].

A.2.2 *Receipts for Plasma Surface Cleaning and Activation*

The specific parameters employed for the cleaning and activation of substrates are expressed in the Table A.2. These values are the result of multiple optimisations of the cleaning process with the goal of achieving the best exfoliation yield.

Table A.2 Plasma cleaning and activation recipes

Equipment	t_{etch} (min)	Φ_g	P_{RF} (W)
GaLa Instrumente Plasma Prep 2	10	98% argon/2% oxygen	20
Oxford Instruments PlasmaPro® NGP80	3	40 SCCM oxygen	50

A.2.3 Derivation of the xSThM Probe Resistance Equation

After having obtained the corrected electrical resistances and powers at both 20 and 60 °C as well as the respective resistance values at zero power, the resistance change as a function of temperature can be written as:

$$\frac{dR_{p,e}}{dT_p} = \frac{\left[R_{p,e}(0)\right]_{60^\circ\mathrm{C}} - \left[R_{p,e}(0)\right]_{20^\circ\mathrm{C}}}{60 - 20[^\circ\mathrm{C}\,]} \tag{A.1}$$

and the change of resistance versus power at 20 °C follows a similar expression.

$$\frac{dR_{p,e}}{dP_p} = \frac{\left(P_p\right)_{20^\circ\mathrm{C}} - 0[W]}{\left[R_{p,e}\left(P_p\right)\right]_{20^\circ\mathrm{C}} - \left[R_{p,e}(0)\right]_{20^\circ\mathrm{C}}} \tag{A.2}$$

Finally, since the thermal resistance is equivalent to the difference of temperature as a function of power, it can be calculated as:

$$R_{p,th} = \frac{dT_p}{dP_p} = \frac{dT_p dR_{p,e}}{dR_{p,e} dP_p} = \frac{40\left\{\left[R_{p,e}\left(P_p\right)\right]_{20^\circ\mathrm{C}} - \left[R_{p,e}(0)\right]_{20^\circ\mathrm{C}}\right\}}{\left\{\left[R_{p,e}(0)\right]_{60^\circ\mathrm{C}} - \left[R_{p,e}(0)\right]_{20^\circ\mathrm{C}}\right\}\left(P_p\right)_{20^\circ\mathrm{C}}} \tag{A.3}$$

Which represents the thermal resistance of the probe at ambient conditions and out of contact.

A.2.4 Ultrasonic Force Microscopy Calibration

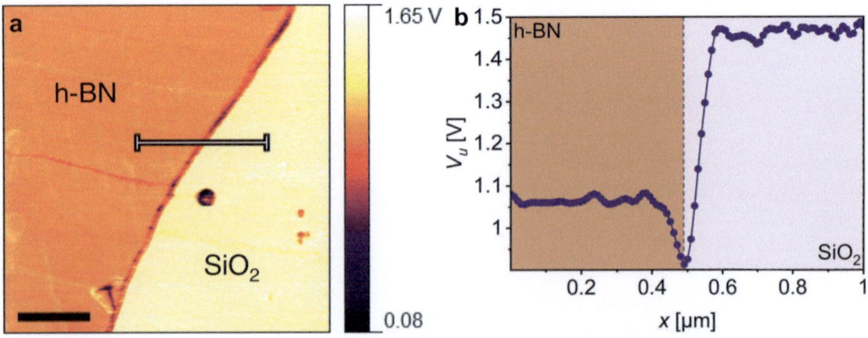

Fig. A.3 UFM calibration for signal conversion. **a** UFM map of an h-BN/SiO_2 sample (scale bar is 500 nm) and **b** corresponding extracted profile of the UFM signal

Before proceeding with the UFM analysis of the TDBG, a calibration must be performed in order to be able to convert the initial UFM signal (V_u) into the Young's modulus parameter (E), assuming the tip's apex to be homogeneous for all the measurements. For this purpose, an area comprising h-BN and SiO_2 (two materials with well-known Young's moduli) of the same sample is measured with the UFM system and shown in Fig. A.3a. Extracting a profile of such data (Fig. A.3b) allows to quantitatively compare the ultrasonic mechanical response of both insulator materials, knowing that their cross-plane (corresponding to the vertical indentation direction) Young's moduli are 38 and 70 GPa for h-BN and SiO_2, respectively [13, 14]. Thus, from the UFM profile and the literature Young's modulus values, one can observe that a 26.89% decrease in V_u ($\Delta V_{u,ref}$) corresponds to a 45.71% decrease in E (ΔE_{ref}). The Young's modulus value of TDBG (E_{TDBG}) is assumed to be similar to FLG, hence, 39.51 GPa [15], and it corresponds to the average of the UFM signal located in the domains. Due to the deep indentation of the UFM probe, the quantification of h-BN must be accounted for, however, the close cross-plane E values of h-BN and TDBG allow a high measurement sensitivity of the TDBG layer. To know the Young's modulus of the domain walls (E_{SP}), the following equation can therefore be applied:

$$E_{SP} = E_{TDBG}\left(1 - \frac{\left(V_{u,max} - V_{u,min}\right)\Delta E_{ref}}{V_{u,max}\Delta V_{u,ref}}\right) \tag{A.4}$$

where $V_{u,min}$ and $V_{u,max}$ are the minimum and maximum UFM signal, respectively. Assuming a linear dependence, the UFM signal can therefore be converted to Young's modulus by:

$$\left.\begin{aligned} E_{TDBG} &= V_{u,max}M + Q \\ E_{SP} &= V_{u,min}M + Q \end{aligned}\right\} \rightarrow \textit{Calculate } M \textit{ and } Q \rightarrow E = V_u M + Q \tag{A.5}$$

A.2.5 Spectroscopic Characterisation

Molybdenum Disulphide/Antimony Telluride Superlattice

The corresponding structural phases of the Sb_2Te_3 and MoS_2 layers are analysed using a HORIBA Semiconductor LabRAM HR Evolution Raman microscope with an argon ion beam excitation source of 514.50 nm of wavelength. Figure A.4a shows the resulting Raman spectra for each of the samples. The active modes of the Sb_2Te_3 layers are visible in the initial range of the spectra. Specifically, they correspond to the $A_{1g}{}^1$, $E_g{}^2$ and $A_{1g}{}^2$ peaks at 60, 114 and 162 cm^{-1}, respectively. The other peaks appearing in the pristine Sb_2Te_3 sample represent the interaction between tellurium atoms or its surface oxide layer response [16]. The in-plane ($E_{2g}{}^1$ at 383 cm^{-1}) and

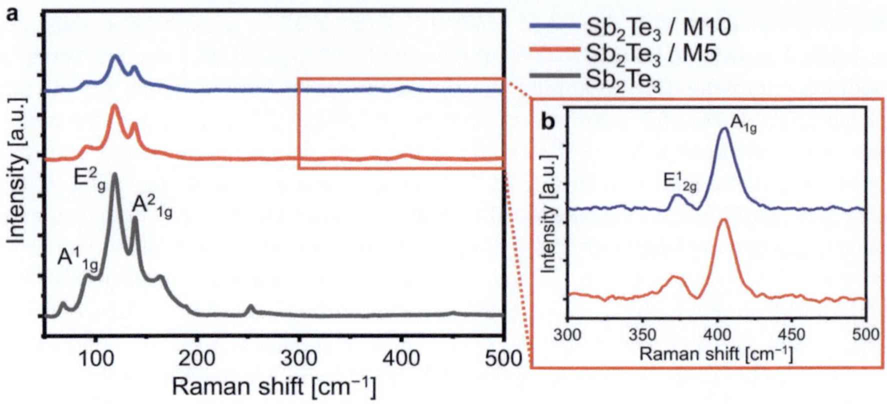

Fig. A.4 **a** Raman spectra of the samples with **b** magnified view of the superlattices' MoS_2 peaks

cross-plane (A_{1g} at 411 cm^{-1}) active Raman modes of the MoS_2 layers [17] are shown in Fig. A.4b, demonstrating the presence of MoS_2 in the multilayer samples together with the Sb_2Te_3 peaks on the left of the spectra. The difference between the two MoS_2 peaks (28 cm^{-1}) indicates that these structures are composed of more than 7–8 monolayers.

Silver/Antimony Telluride Nanocomposite

To characterise the layers' composition and corroborate the formation of distinct phases in the nanocomposites, X-ray diffraction (XRD) and RT Raman spectroscopy measurements are conducted. Employing a PANalytical X'Pert PRO X-ray diffractometer with a copper Kα irradiation of 0.15 nm of wavelength, one can extract the

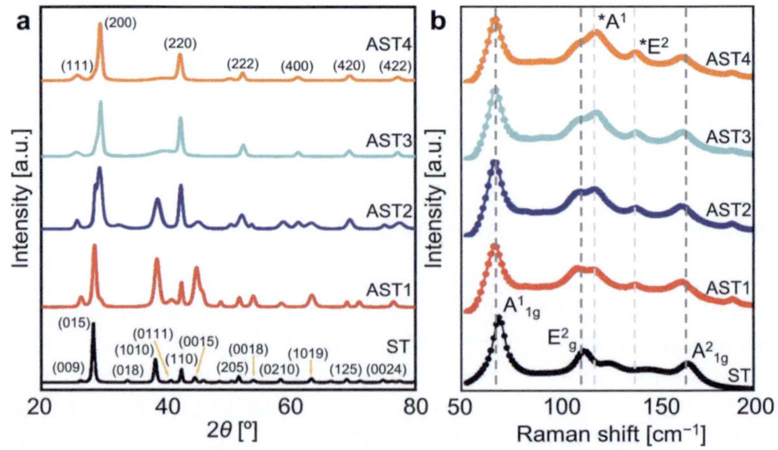

Fig. A.5 **a** XRD and **b** Raman spectra of the nanocomposite samples

diffraction patterns represented in Fig. A.5a. The typical trigonal response of Sb_2Te_3 can be seen in all the samples at the (009), (015), (1010), (0015) and (0210) peaks. When the silver concentration is increased, a new (200) peak appears, which corresponds to the lattice plane of the $AgSbTe_2$ phase. The consecutive weakening of the Sb_2Te_3 peaks with increased silver concentration is attributed to the loss of crystallinity of the structure, given that silver doping rises the crystallisation temperature and decreases the crystalline activation energy [18]. The Raman study is performed with a Renishaw® inVia™ confocal Raman microscope with an excitation source of 514 nm of wavelength and low power of < 1 mW to avoid sample damage. Figure A.5b shows the obtained Raman spectra of the five analysed samples, where the dark grey dashed lines indicate the position of the Raman modes corresponding to Sb_2Te_3 ($A_{1g}{}^1$, $E_g{}^2$ and $A_{1g}{}^2$ peaks at 68.70, 111.90 and 165.50 cm^{-1}, respectively) and the light grey dashed lines mark the redshift of such modes caused by local bonding environment variation in the silver nanocomposites.

Boron Nitride–Carbon Nanotubes

Composition characterisation is undertaken by electron energy loss spectroscopy (EELS) measurements (Fig. A.6a–c) to reveal further evidence of the h-BN encapsulation of MWCNT. Figure A.6a and c show prominent π^* and σ^* antibonding peaks on the graphene (C-K) and h-BN (B-K and N-K) structures of the coaxial nanotubes, arising from the empty π^* and σ^* orbitals to which excited 1 s core

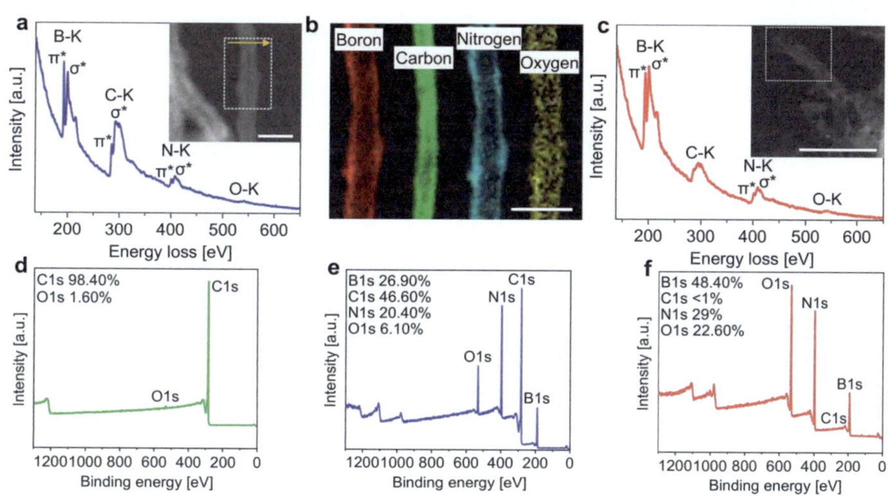

Fig. A.6 Composition characterisation of the nanotubes. **a** EELS spectrum of MWCNT@BNNT (inset shows the STEM image used to extract the profile along the orange arrow) (scale bar is 20 nm), **b** EELS spectral images of the elements present in the rectangle of the inset in **a** (scale bar is 20 nm), **c** EELS spectrum of a MWBNNT (inset shows the STEM image used to extract the profile) (scale bar is 100 nm) and **d–f** XPS spectra of the MWCNT, MWCNT@BNNT and MWBNNT BPs, respectively

electrons are promoted when ionisation is produced. However, the C-K peak in MWBNNT (Fig. A.6c) is less pronounced due to its lack of carbonic structures. The polymorphism of the coaxial nanotube is confirmed by observing its lamellar nature in Fig. A.6b, where the positions of each element in the MWCNT@BNNT are represented. This further depicts an outer shell of boron and nitrogen covering the carbon nanotube. Oxygen positions show a diminished intensity compared to the other elements, nevertheless, their superimposition matches with the outer h-BN shell, suggesting an oxidation of such layer. The presence of a carbon peak in Fig. A.6c could mean an incomplete removal of the MWCNT cores during the thermal oxidation process, however, a close inspection of such peak indicates its resemblance to the amorphous carbon response [19], containing π^* and σ^* peaks of lower intensity. The appearance of such amorphous carbon is due to adventitious carbon contamination prevalent on surfaces exposed to air [19, 20].

XPS measurements are also conducted in each sample, with the extracted spectra shown in Fig. A.6d–f. The C1s peaks appearing in Fig. A.6d and e (corresponding to the MWCNT and MWCNT@BNNT BPs, respectively) are quasi-identical, with their difference not showing signal variations in the carbon/boron or carbon/nitrogen regions, meaning that the CNT core is maintained intact and without appreciable covalent binding with boron or nitrogen after the coaxial fabrication. This is validated by a close inspection of the B1s and N1s peaks, showing a predominant bonding structure between boron and nitrogen. Interestingly, Fig. A.6f shows a minimal carbon response in MWBNNT BP characteristic of amorphous carbon, corroborating the previous EELS explanation. Also, the XPS spectra of the MWCNT@BNNT and MWBNNT BPs show high oxygen peaks, confirming the presence of oxidised functional groups in the h-BN structures. Said oxidation could be due to contamination on the CVD tube, given that atmospheric pressure CVD is open to airflow between uses and there is no possible vacuum pumping precleaning procedure, leaving residual atmospheric gases after purging [21]. However, the presence of a hydrogen flux during the CVD process is to minimise oxygen impurities, decreasing the total parasitic passivation [21]. Another source of contamination could be the exposure of the BPs to an ambient atmosphere and moisture during the sample storage, causing a hydraulic attack on h-BN [22].

High resolution TEM measurements show an imperfect crystallinity at both the CNT core and the BNNT shell, increasing the defect quantity, which could be a source for water physisorption (oxidation) in h-BN [23]. Thermal exposure of materials without high crystallinity can also increase oxidation, thus an intensity enhancement of the O1s peaks of 16.50% is observed when the MWCNT@BNNT BP is thermally oxidised to form the MWBNNT BP.

A.3 Nanocomposite xSThM Preliminary Data

As explained in Sect. 5.3.2, topography and thermal signal data are obtained from xSThM and shown in Fig. A.7. Similar to the previously commented structures, the topographical data does not show a clear transition between the materials present in the sample, nevertheless, due to their different thermal properties, it is possible to differentiate such layers on the thermal signal images and profiles, simultaneously scanned with the topography. This simultaneity allows to compute both the topographical and thermal data at the same position and combine them in Eq. (4.6) to express the experimental results as thermal resistance as a function of thickness.

The thermal data extracted from the structures can also be compared to the "Muzychka-Spièce" analytical modelling to quantify the individual anisotropic thermal conductivities and interfacial thermal insulances of the nanocomposite samples. Therefore, to delimitate the ranges of such parameters, Fig. A.8 is employed after obtaining the contact radius and correction factor from the reference samples' data fitting (using the same literature data explained in Sect. 5.2.3). These maximum and minimum values for the thermal conductivities are employed to improve the quality of the anisotropic fitting of the experimental data for each top layer thin film, disregarding the pristine sample given its higher thermal conductivity compared to the substrate, thus, not allowing the use of the theoretical model (see Sect. 5.2.3).

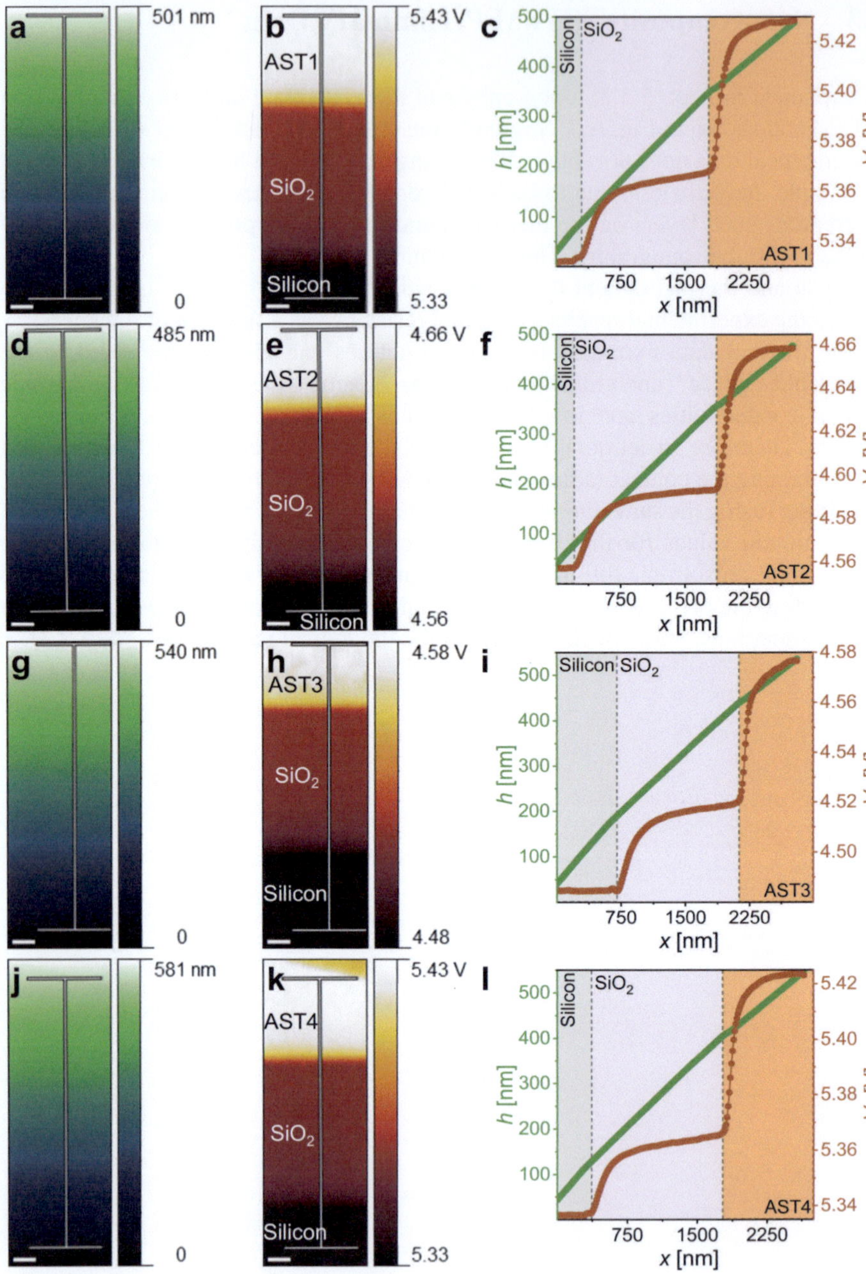

Fig. A.7 xSThM maps and profiles of the analysed samples. **a**, **d**, **g**, **j** Topography maps, **b**, **e**, **h**, **k** thermal signal maps (scale bars are 200 nm) and **c**, **f**, **i**, **l** corresponding profiles of the AST1, AST2, AST3 and AST4 samples, respectively

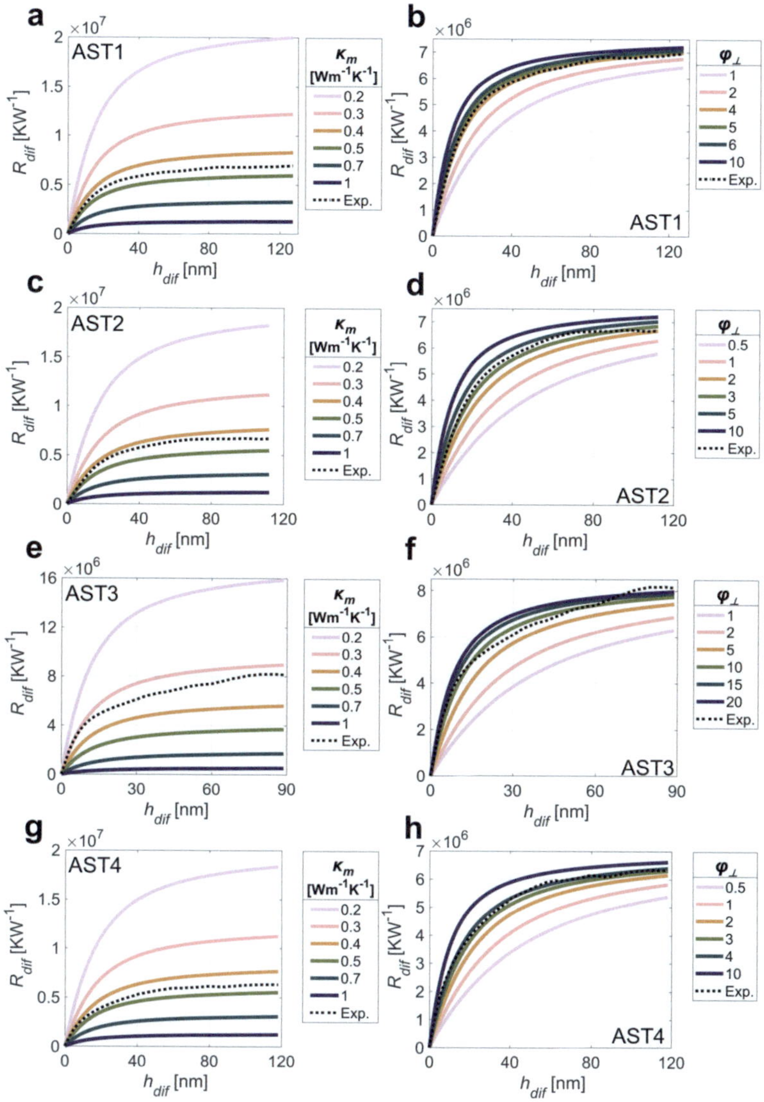

Fig. A.8 Comparison of simulated $R_{dif}(h_{dif})$ dependences according to the analytical model (coloured lines) versus the experimental data (black dotted lines) for the anisotropic parameters of the nanocomposites. **a**, **c**, **e**, **g** Varying κ_m and **b**, **d**, **f**, **h** varying $\varphi_\perp$

A.4 Low Temperature Thermovoltage Maps of the Hole Arrays

Liquid nitrogen is employed to cool the sample at LT (~ 200 K) to increase the thermovoltage response after the temperature gradient between the tip and the electrodes increases.

The patterns with different pitches are firstly measured following the stipulated STGM procedure, obtaining a better resolution due to the enhanced thermovoltage (Fig. A.9). A destructive interference phenomenon is also observed at low temperatures as the pitch between the holes decreases, together with the side thermovoltage gradients explained before. However, the thermovoltage response is more uniform for each of the circular constrictions compared to the RT measurements. The changes in thermovoltage polarity for such constrictions have lower amplitudes as the pitch decreases, following the RT trend, but the side gradients seem to have similar amplitude, suggesting that, at low temperatures, the length of the patterned rectangles only affects the individual constriction response rather than the polygonal TE. Outside of the patterns, other topographical features are more clearly reflected in the TE signal, given the enhanced temperature difference creating charge carriers with higher energy able to scatter on more topographic defects.

The same measurements are conducted for the patterns with different hole quantities (Fig. A.10), showing an enhanced thermovoltage due to the decrease in the sample's temperature. The same phenomena happening at RT are observed here,

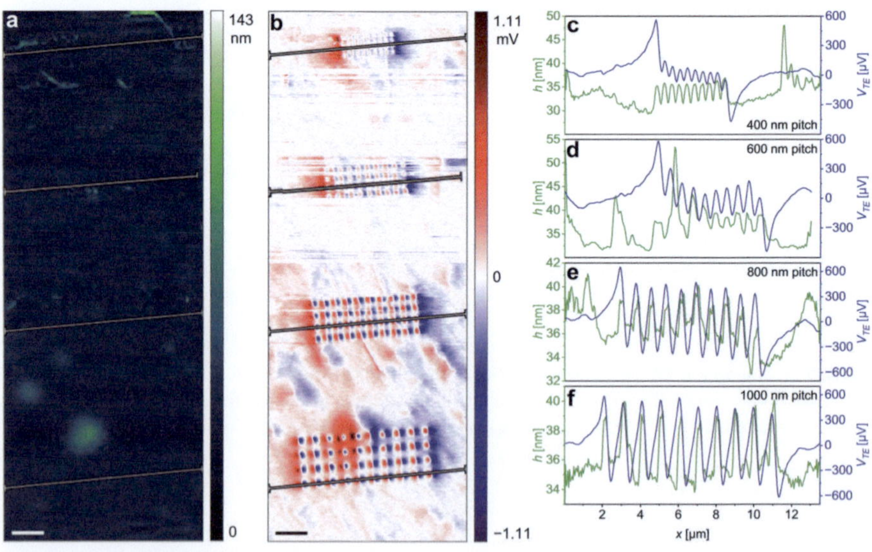

Fig. A.9 LT STGM analysis of the different pitch hole arrays. **a** Topography and **b** thermovoltage maps of the patterns (scale bars are 2 μm) and **c**, **d**, **e** and **f** corresponding extracted profiles for the matrices with 400, 600, 800, 1000 nm of pitch, respectively

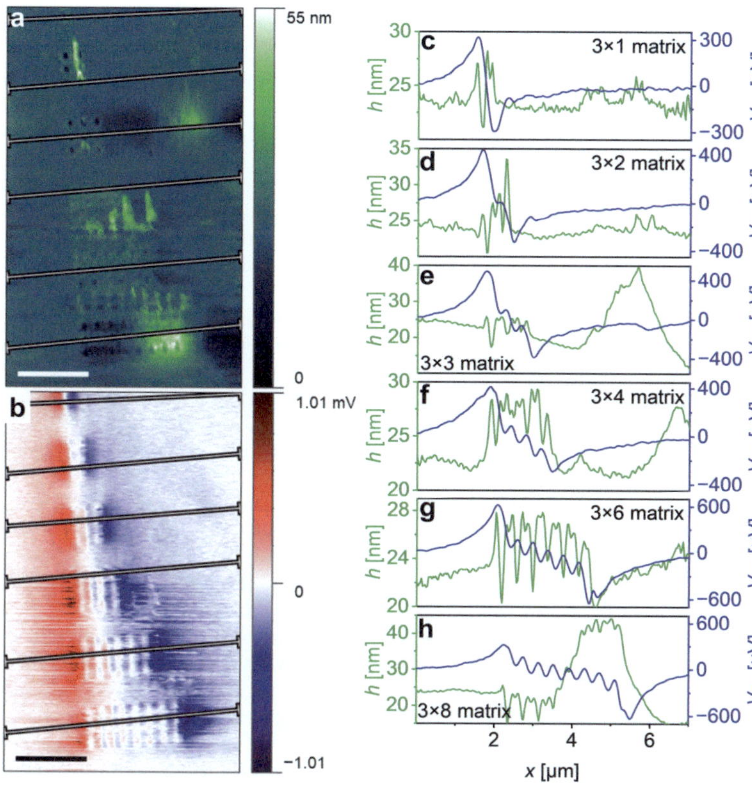

Fig. A.10 LT STGM analysis of the arrays with different number of holes. **a** Topography and **b** thermovoltage maps of the patterns (scale bars are 2 μm) and **c**, **d**, **e**, **f**, **g** and **h** corresponding extracted profiles for the 3-row matrices with 1, 2, 3, 4, 6 and 8 columns of holes, respectively

including the tilted slope due to the close proximity of the rectangles' side thermoelectric influence. Nevertheless, the amplitudes of such side thermovoltages drastically vary without following any clear trend. This could be due to the more prominent influence of other topographical parasitic phenomena as observed at the brightest spots in the topography map.

References

1. Gao, E. et al. (2018). Mechanical exfoliation of two-dimensional materials. *Journal of the Mechanics and Physics of Solids, 115*, 248–262.
2. Tang, T., Wang, Z., & Guan, J. (2022). A review of defect engineering in two-dimensional materials for electrocatalytic hydrogen evolution reaction. *Chinese Journal of Catalysis, 43*, 636–678.

3. Huang, Y., et al. (2020). Universal mechanical exfoliation of large-area 2D crystals. *Nature Communications, 11*, 2453.
4. Imani Yengejeh, S., Liu, J., Kazemi, S. A., Wen, W., & Wang, Y. (2020). Effect of structural phases on mechanical properties of molybdenum disulfide. *ACS Omega, 5*, 5994–6002.
5. Li, X., et al. (2021). Tunable negative Poisson's ratio in van der Waals superlattice. *Research, 2021*, 1–11.
6. Liu, X., Metcalf, T. H., Robinson, J. T., Houston, B. H., & Scarpa, F. (2012). Shear modulus of monolayer graphene prepared by chemical vapor deposition. *Nano Letters, 12*, 1013–1017.
7. Zheng, Q., Liang, C., Jiang, J., & Li, S. (2022). Elastic properties and deformation mechanisms in the van der Waals single-crystalline indium selenide. *Physica Status Solidi RRL, 16*, 2100418.
8. Tamulewicz-Szwajkowska, M., Zelewski, S. J., Serafińczuk, J., & Kudrawiec, R. (2022) Geometric progress in the thickness of exfoliated van der Waals crystals on the example of MoS_2. *AIP Advances, 12*, 025328.
9. Mag-isa, A. E., Kim, J.-H., Lee, H.-J., & Oh, C.-S. (2015). A systematic exfoliation technique for isolating large and pristine samples of 2D materials. *2D Materials, 2*, 034017.
10. Huang, Y., et al. (2015). Reliable exfoliation of large-area high-quality flakes of graphene and other two-dimensional materials. *ACS Nano, 9*, 10612–10620.
11. Thompson, J. P., Doha, M. H., Murphy, P., Hu, J., & Churchill, H. O. H. (2019). Exfoliation and analysis of large-area, air-sensitive two-dimensional materials. *Journal of Visualized Experiments, 143*, e58693.
12. Wang, J. I. J., et al. (2015). Electronic transport of encapsulated graphene and WSe_2 devices fabricated by pick-up of prepatterned hBN. *Nano Letters, 15*, 1898–1903.
13. Carlotti, G., Doucet, L., & Dupeux, M. (1997). Elastic properties of silicon dioxide films deposited by chemical vapour deposition from tetraethylorthosilicate. *Thin Solid Films, 296*, 102–105.
14. Woo, S., Park, H. C., & Son, Y.-W. (2016). Poisson's ratio in layered two-dimensional crystals. *Physical Review B, 93*, 075420.
15. Cho, J., Luo, J. J., & Daniel, I. M. (2007). Mechanical characterization of graphite/epoxy nanocomposites by multi-scale analysis. *Composites Science and Technology, 67*, 2399–2407.
16. Ahmad, M., Varandani, D., & Mehta, B. R. (2018). Large surface charge accumulation in 2D MoS_2/Sb_2Te_3 junction and its effect on junction properties: KPFM based study. *Applied Physics Letters, 113*, 141603.
17. Bagnall, A. G., Liang, W. Y., Marseglia, E. A., & Welber, B. (1980). Raman studies of MoS_2 at high pressure. *Physica B + C, 99*, 343–346.
18. Xu, J., Liu, B., Song, Z., Feng, S., & Chen, B. (2006). Crystallization and C-RAM application of Ag-doped Sb_2Te_3 material. *Materials Science and Engineering B, 127*, 228–232.
19. Kim, K. S., et al. (2014). Hydrogen-catalyzed, pilot-scale production of small-diameter boron nitride nanotubes and their macroscopic assemblies. *ACS Nano, 8*, 6211–6220.
20. Miller, D. J., Biesinger, M. C. & McIntyre, N. S. Interactions of CO_2 and CO at fractional atmosphere pressures with iron and iron oxide surfaces: one possible mechanism for surface contamination? *Surf. Interface Anal.* **33**, 299–305 (2002).
21. Reckinger, N., Felten, A., Santos, C. N., Hackens, B., & Colomer, J.-F. (2013). The influence of residual oxidizing impurities on the synthesis of graphene by atmospheric pressure chemical vapor deposition. *Carbon, 63*, 84–91.
22. Matsuda, T. (1989). Stability to moisture for chemically vapour-deposited boron nitride. *Journal of Materials Science, 24*, 2353–2357.
23. Anutgan, T. A., Anutgan, M., Atilgan, I., & Katircioglu, B. (2009). Stability and degradation of plasma deposited boron nitride thin films in ambient atmosphere. *Thin Solid Films, 518*, 419–425.

Curriculum Vitae: Sergio Gonzalez-Munoz

Summary

I am a postdoctoral research fellow in the School of Materials Science and Engineering at the Georgia Institute of Technology, conducting research in nanoscale characterisation of the optomechanical properties of polymeric nanomaterials via advanced scanning probe microscopy methodologies. My research is funded by the U.S. National Science Foundation and the U.S. Air Force Research Laboratory.

The results of my research include the measurement of chiroptical properties of mechanically enhanced cellulose materials, the quantification of the nanothermal anisotropy of vdW materials, the improved thermoelectric behaviour of doped 2D heterostructures and the mapping of strain in electromechanical Moiré-like patterns formed in twisted double bilayer graphene. My studies have been published in leading journals such as Advanced Functional Materials and Nature Communications and have been selected as the Editor's Choice and Front Cover on multiple occasions. In 2023, I received the E-MRS Young Researcher Award and the Lancaster University Prize for outstanding PhD research and contributions to high-impact publications.

S. Gonzalez-Munoz, *Interplay of Nanoscale Physical Properties in 2D Materials and Heterostructures*, Springer Theses, https://doi.org/10.1007/978-3-032-14271-9

Education

• PhD in physics 2020–2024

Physics Department, Lancaster University, United Kingdom

Thesis: Interplay of nanoscale physical properties in 2D materials and heterostructures.

• MSc in advanced nanoscience and nanotechnology 2019–2020

Faculty of Sciences, Autonomous University of Barcelona, Spain

Thesis: Reduced graphene oxide thin films for neural interfaces.

• BEng in mechanical engineering 2014–2018

Higher Polytechnic School, University of Lleida, Spain

Thesis: "Save Bambi" project in robotics.

Awards/Grants

- Young researcher award. *E-MRS 2023 Fall Meeting, European Materials Research Society* (2023).
- Lancaster University departmental prize for outstanding PhD research and contributions to high-impact publications. *Physics Department, Lancaster University* (2023).
- Graphene Week student grant. *Graphene Flagship, European Union* (2023).
- Research student conference fund. *Institute of Physics, United Kingdom* (2023).
- Graphene Week student grant. *Graphene Flagship, European Union* (2022).
- Postgraduate scholarship. *Lancaster University* (2020–2024).
- Scholarship for Erasmus + studies (Denmark, EU). *European Commission* (2018).

Research Experience

- Materials Science and Engineering, Georgia Institute of Technology — 2024–Present

Postdoctoral research fellow — Supervisor: Prof. Vladimir V. Tsukruk

Project: "Biophotonic properties of natural polymers and their chiral assembly"—Funded by the U.S. National Science Foundation and the U.S. Air Force Research Laboratory.

- Spectrophotometry of Bouligand cellulose nanocrystal structures.
- HR-AFM and microindentation of topographic and mechanical properties.
- FDTD/FEM simulations of gradient CNC/MXene structures.

- Physics Department, Lancaster University — 2020–2024

Postgraduate researcher — Supervisor: Prof. Oleg V. Kolosov

Project: "Nanoscale transport and quantum mechanical phenomena in 2D materials nanostructures"—Funded by the Graphene Flagship, European Union.

- Analysis of anisotropic thermal properties of vdW materials.
- Electromechanical SPM of "Moiré-like" patterns.
- Fabrication of 2D samples, heterostructures and devices.

- Physics Department, Lancaster University — 2023

Research associate — Supervisors: Prof. Oleg V. Kolosov & Prof. Nuria Tapia Ruiz

Project: "NEXGENNa"—Funded by the European Union and the UKRI.

– Characterisation of solid-electrolyte interphases in batteries using SPM.

- Physics Department, Lancaster University — 2020–2021

Research associate — Supervisor: Prof. Oleg V. Kolosov

Project: "CS Hub"—Funded by the EPSRC, UKRI.

– Fabrication and SPM characterisation of compound semiconductors.

- Catalan Institute of Nanoscience and Nanotechnology (ICN2) 2019–2020

Visiting researcher Supervisor: Prof. Jose A. Garrido

Project: "Graphene thin films for neural interfaces".

- Fabrication of neural devices based on graphene interfaces.

Teaching Experience

- Materials Science and Engineering, Georgia Institute of Technology 2024–2025

Lecturer substitute

- Polymer Characterisation (MSE/CHEM/CHBE 6752)—Lecture teaching, 14 PhD postgraduate year 1 students from different Materials Science and Chemistry courses.
- Soft Nano Bio Materials (MSE 4335)—Lecture teaching, 27 postgraduate and undergraduate students from different Materials Science and Engineering courses.

- Physics Department, Lancaster University 2020–2022

Postgraduate teaching assistant

- Physics Skills I/Practical Physics I (PHYS131)—Marking and workshop teaching, 172 undergraduate year 1 students from different physics courses.
- Thermal Properties of Matter (PHYS233)—Marking and workshop teaching, 143 undergraduate year 1 and year 2 students from different physics courses.

Research Skills

- *Scanning probe microscopy (SPM) techniques:* AFM (contact, tapping and peak force modes), SThM, PFM, UFM, QNM, STGM, KPFM, SSRM and E-HFM. Environments: air, liquid, high-vacuum, Ar and low temperature.
- *Other characterisation techniques:* Optical microscopy, SEM, Raman spectroscopy, LDV, spectrophotometry, tensile testing, nanoindentation and profilometry.

- *Fabrication techniques:* Exfoliation, stacking, spin coating, blade coating, vacuum filtering, optical lithography, SPM patterning, metal deposition, RIE and BEXP.
- *Special working environments:* Glovebox and cleanroom operation.
- *IT skills:* MATLAB, Gwyddion, OriginLab, AutoCAD, PTC Creo, COMSOL, Ansys, LabVIEW and ImageJ.

Affiliations

- Member of the Institute of Physics (IOP), 2022–2023.
- Member of the Materials Research Society (MRS), 2021.

Publications

- Mech, R., Canetta, A., Huang, Y., Gonzalez-Munoz, S., Agarwal, K., de Crombrugghe, P., Hong, Y., Mohapatra, S., Watanabe, K., Taniguchi, T., Nysten, B., Hackens, B., Ribeiro Palau, R., Kolosov, O., Spiece, J. & Gehring, P. Thermal transport mapping in twisted double bilayer graphene (Under review in *Appl. Phys. Lett.*).
- Medapalli, R., Cottam, N. D., Agarwal, K., Dewes, B. T., Dessmann, N., Gonzalez-Munoz, S., Yan, W., Mišeikis, V., Kafanov, S., Mikhaylovskiy, R. V., Jarvis, S., Coletti, C., Redlich, B., Patanè, A. & Kolosov, O. V. Broadband diamond campanile probe enables efficient mid-infrared light concentration and sub-wavelength photocurrent imaging in graphene (Under review in *Nat. Photonics*).
- Brackenridge, J., Gonzalez-Munoz, S., Bukharina, D., Poliukhova, V. & Tsukruk, V. V. Double Bouligand printed twisted stacks: mirrored chiroptical polarization and enhanced compressive mechanics (Under review in *Adv. Opt. Mater.*).
- Ghosh, A., Agarwal, K., Gonzalez Munoz, S., Bisht, P., Vishwakarma, C. K., Kaur, N., Ahmad, M., Vullum, P. E., Belle, B. D., Singh, R., Kolosov, O. V. & Mehta, B. R. Enhancing thermoelectric properties of Sb_2Te_3-$AgSbTe_2$ nanocomposite thin film - synergy of band engineering and heat transport modulation (Under review in *Adv. Sci.*).
- Kang, D., Kim, J., Gonzalez Munoz, S., Jeon, J., Park, S. & Tsukruk, V. V. Piezoionic materials and structures for complex shear field monitoring. *ACS Appl. Mater. Interfaces* **17**, 41350–41360 (2025).
- Canetta, A., Gonzalez-Munoz, S., Nguyen, V.-H., Agarwal, K., de Crombrugghe de Picquendaele, P., Hong, Y., Mohapatra, S., Watanabe, K., Taniguchi, T., Nysten, B., Hackens, B., Ribeiro-Palau, R., Charlier, J.-C., Kolosov, O. V., Spièce, J. & Gehring, P. Quantifying the local mechanical properties of twisted double bilayer graphene. Preprint at https://arxiv.org/abs/2405.12610 (2024).

- Jones, R. S., Gonzalez-Munoz, S., Griffiths, I., Holdway, P., Evers, K., Luanwuthi, S., Maciejewska, B. M., Kolosov, O. & Grobert, N. Thermal conductivity of carbon/boron nitride heteronanotube and boron nitride nanotube buckypapers: implications for thermal management composites. *ACS Appl. Nano Mater.* **6**, 15374–15384 (2023). **Highlighted as a Front Cover**.
- Gonzalez-Munoz, S., Agarwal, K., Castanon, E. G., Kudrynskyi, Z. R., Kovalyuk, Z. D., Spièce, J., Kazakova, O., Patanè, A. & Kolosov, O. V. Direct measurements of anisotropic thermal transport in γ-InSe nanolayers via cross-sectional scanning thermal microscopy. *Adv. Mater. Interfaces* **10**, 2300081 (2023). **Highlighted as a Front Cover and Editor's Choice**.
- Ghosh, A., Agarwal, K., Gonzalez Munoz, S., Bisht, P., Vishwakarma, C. K., Kaur, N., Ahmad, M., Vullum, P. E., Belle, B. D., Singh, R., Kolosov, O. V. & Mehta, B. R. Tuning thermoelectric properties of Sb_2Te_3-$AgSbTe_2$ nanocomposite thin film - synergy of band engineering and heat transport modulation. Preprint at https://arxiv.org/abs/2305.19217 (2023).
- Canetta, A., Gonzalez-Munoz, S., Nguyen, V.-H., Agarwal, K., de Crombrugghe de Picquendaele, P., Hong, Y., Mohapatra, S., Watanabe, K., Taniguchi, T., Nysten, B., Hackens, B., Ribeiro-Palau, R., Charlier, J.-C., Kolosov, O. V., Spièce, J. & Gehring, P. Quantifying the local mechanical properties of twisted double bilayer graphene. *Nanoscale* **15**, 8134–8140 (2023). **Highlighted as an Inside Front Cover**.
- Chen, Y., Wu, W., Gonzalez-Munoz, S., Forcieri, L., Wells, C., Jarvis, S. P., Wu, F., Young, R., Dey, A., Isaacs, M., Nagarathinam, M., Palgrave, R. G., Tapia-Ruiz, N. & Kolosov, O. V. Nanoarchitecture factors of solid electrolyte interphase formation via 3D nano-rheology microscopy and surface force-distance spectroscopy. *Nat. Commun.* **14**, 1321 (2023).
- Maiti, A., Agarwal, K., Gonzalez-Munoz, S. & Kolosov, O. V. Quantifying anisotropic thermal transport in two-dimensional perovskite (PEA_2PbI_4) through cross-sectional scanning thermal microscopy. *Phys. Rev. Mater.* **7**, 023801 (2023).
- Ahmad, M., Agarwal, K., Gonzalez Munoz, S., Ghosh, A., Kodan, N., Kolosov, O. V. & Mehta, B. R. Engineering interfacial effects in electron and phonon transport of Sb_2Te_3/MoS_2 multilayer for thermoelectric ZT above 2.0. *Adv. Funct. Mater.* **32**, 2206384 (2022).

Conference Participation

- Oral presentations
  - Probing the nanophysics interplay via advanced scanning probe microscopy. *Ludwig Maximilian University of Munich 2024* (Munich, Germany).
  - Quantitative probing of nanoscale heat transport via advanced scanning thermal microscopy. *National Institute of Standards and Technology 2024* (Virtual).

- Quantitative scanning thermal microscopy studies of the influence of interfaces and heat transport anisotropy in 2D materials. *E-MRS 2023 Fall Meeting* (Warsaw, Poland).
- Thermomechanical study via piercing indentation SThM in C/BN hetero-nanotubes buckypapers. *Microscience Microscopy Congress (MMC) 2023* (Manchester, United Kingdom).
- Measurements of anisotropic thermal transport in γ-InSe via quantitative cross-sectional scanning thermal microscopy. *Virtual RMS AFM & SPM Meeting 2022* (Virtual).
- Quantitative measurements of anisotropic thermal transport in gamma indium selenide via cross-sectional scanning thermal microscopy. *Condensed Matter and Quantum Materials (CMQM) 2022* (Bath, United Kingdom).
- Quantitative measurements of anisotropic thermal transport in vdW materials via cross-sectional scanning thermal microscopy (xSThM). *E-MRS 2022 Spring Meeting* (Virtual).
- Quantitative measurements of anisotropic thermal transport in vdW materials via cross-sectional scanning thermal microscopy (xSThM). *2022 MRS Spring Meeting & Exhibit* (Honolulu, United States).

- Posters
  - Gonzalez-Munoz, S., Agarwal, K. & Kolosov, O. V. Enhanced thermoelectricity and thermal anisotropy of 2D-based structures. *AFM & SPM Meeting 2024* (Durham, United Kingdom).
  - Gonzalez-Munoz, S., Agarwal, K., Maiti, A. & Kolosov, O. V. Quantitative scanning thermal microscopy studies of the influence of interfaces and heat transport anisotropy in 2D materials. *Graphene Week 2023* (Gothenburg, Sweden).
  - Gonzalez-Munoz, S., Canetta, A., Agarwal, K., Spièce, J., Gehring, P. & Kolosov, O. V. Correlation of nanoscale electromechanical and mechanical properties of twisted double bi-layer graphene via UFM, PFM, and E-HFM. *Microscience Microscopy Congress (MMC) 2023* (Manchester, United Kingdom).
  - Gonzalez-Munoz, S., Agarwal, K., Maiti, A. & Kolosov, O. V. Quantitative scanning thermal microscopy studies of the influence of interfaces and heat transport anisotropy in 2D materials. *Microscience Microscopy Congress (MMC) 2023* (Manchester, United Kingdom).
  - Gonzalez-Munoz, S., Agarwal, K., Ghosh, A., Ahmad, M., Mehta, B. R. & Kolosov, O. V. Anisotropic thermal transport using xSThM studies in 2D-3D heterostructures and composites. *Graphene Week 2022* (Munich, Germany).
  - Gonzalez-Munoz, S., Agarwal, K., Castanon, E. G., Kudrynskyi, Z. R., Kovalyuk, Z. D., Spièce, J., Kazakova, O. L., Patanè, A. & Kolosov, O. V. Measurements of anisotropic thermal transport in γ-InSe via quantitative cross-sectional scanning thermal microscopy. *Graphene Week 2022* (Munich, Germany).

- Gonzalez-Munoz, S., Agarwal, K., Castanon, E., Kudrynskyi, Z., Kovalyuk, Z., Spièce, J., Kazakova, O., Patanè, A. & Kolosov, O. Quantitative measurements of anisotropic thermal transport in gamma indium selenide via cross-sectional scanning thermal microscopy. *Faculty of Science and Technology (FST) Science Week 2022* (Lancaster, United Kingdom).
- Gonzalez-Munoz, S., Agarwal, K., Castanon, E., Kudrynskyi, Z., Kovalyuk, Z. D., Kazakova, O., Patanè, A. & Kolosov, O. Measurements of nanoscale thermal transport and its anisotropy in vdW materials via cross-sectional scanning thermal microscopy (xSThM). *Graphene Week 2021* (Virtual).
- Agarwal, K., Gonzalez-Munoz, S., Castanon, E., Niblett, A., Kudrynskyi, Z. R., Kazakova, O., Patanè, A. & Kolosov, O. V. Scanning thermal microscopy of 2D materials in high vacuum environment. *Microscience Microscopy Congress (MMC) 2021* (Virtual).

- Judge panel
  - *Fall 2024 poster showcase and 3-min research challenge*, Georgia Institute of Technology School of Materials Science and Engineering (GT MSE) (Atlanta, United States).
  - *Career, Research, and Innovation Development Conference (CRIDC) 2025*, Georgia Institute of Technology Graduate Student Government Association (Atlanta, United States).
  - *Spring 2025 Poster Showcase*, Georgia Institute of Technology School of Materials Science and Engineering (GT MSE) (Atlanta, United States).

External Administrative Roles

- Peer reviewer for Adv. Mater., Nat. Commun. and ACS Appl. Mater. Interfaces.

MIX
Papier aus verantwortungsvollen Quellen
Paper from responsible sources
FSC® C105338

If you have any concerns about our products,
you can contact us on
ProductSafety@springernature.com

In case Publisher is established outside the EU,
the EU authorized representative is:
Springer Nature Customer Service Center GmbH
Europaplatz 3, 69115 Heidelberg, Germany

Printed by Libri Plureos GmbH
in Hamburg, Germany